Haar Series and Linear Operators

Mathematics and Its Applications

Volume 367

Haar Series and Linear Operators

by

Igor Novikov

and

Evgenij Semenov

Department of Mathematics,
Voronezh State University,
Voronezh, Russia

SPRINGER-SCIENCE+BUSINESS MEDIA, B.V.

Library of Congress Cataloging-in-Publication Data

Novikov, Igor.
Haar series and linear operators / by Igor Novikov and Evgenij Semenov.
p. cm. -- (Mathematics and its applications ; v. 367)
Includes bibliographical references and index.
ISBN 978-90-481-4693-2 ISBN 978-94-017-1726-7 (eBook)
DOI 10.1007/978-94-017-1726-7
1. Haar system (Mathematics) 2. Linear operators. I. Semenov, E. M. II. Title. III. Series: Mathematics and its applications (Kluwer Academic Publishers) ; v. 367.
QA404.5.N68 1996
515'.2433--dc20 96-13618

ISBN 978-90-481-4693-2

Printed on acid-free paper

Table of Contents

Preface

In 1909 Alfred Haar introduced into analysis a remarkable system χ which bears his name. He established the first basic results about this system: χ is a complete orthonormal system on [0,1]; if $x \in C[0,1]$ ($x \in L_1[0,1]$, then the Fourier series of x with respect to χ converges uniformly (almost everywhere) to x on [0,1]. The Haar system was the first one to possess the last two properties. Later it has been found out that χ has other remarkable properties.

J. Schauder proved that χ is a basis in L_p ($1 \leq p \leq \infty$). In fact, χ forms a basis in L_1 in contrast to the trigonometric system. R. Paley and J. Marcinkiewich showed that χ is an unconditional basis in L_p ($1 < p < \infty$). Moreover, if an unconditional basis exists in a rearrangement invariant space E, then χ is an unconditional one in E too, and its unconditional basis constant is minimal. This theorem was proved by A. Olevskii. The Haar system is universal in the martingale theory. By B. Mayrey's theorem, each martingale-difference sequence is equivalent to the block-basis with respect to χ.

Some linear operators are closely connected with χ. Their investigation started simultaneously with the development of the general operator theory. The fundamental results about χ mentioned above have natural operator formulations. This monograph is devoted to the investigation of the Haar system from the point of operator theory.

Let us present the contents of the monograph in brief.

Chapter 1 contains auxiliary results about rearrangement invariant spaces, bases, interpolation methods.

Partial sums and their convergence in the space L_p, $L_p(\omega)$, and in rearrangement invariant ones are studied in Chapter 2-4.

Classical results on unconditional convergence of the Haar series in modern presentation are given in Chapters 5, 6. D. Burkholder has found an exact estimate of norms in L_p ($1 < p < \infty$) of multipliers with respect to χ corresponding a given sequence of ± 1. Such operators in L_1 and L_∞ can be unbounded. By S. Yano's theorem, they are weak type (1,1). An exact estimate of the weak (1,1)-norms of such multipliers has been found by D. Burkholder. He has also proved that the Paley operator has weak type (1,1) and is bounded in L_p ($1 < p < \infty$). A generalization of the B. Davis theorem about the norm equivalence of the Paley function and the majorant of partial sums is also studied in Chapter 6.

We investigated Fourier-Haar coefficients of functions from some functional spaces in Chapter 7.

The correlation between the Haar system and martingales can be found in Chapter 8

The Haar system is reproducible. The results of A. Olevskii and G. Schechtman in this direction are presented in Chapter 9.

The Haar functions are closely connected with monotone bases in rearrangement invariant spaces. A generalization of T. Ando's theorem about contractive projections in L_p and L. Dor - E. Odell theorem about monotone bases in L_p are studied in Chapter 10.

We investigate the rearrangements of χ in L_p, $L_{p,q}$ and BMO spaces in Chapter 11 and multipliers with respect to χ in Chapters 12-14. The necessary and sufficient conditions for boundedness of the multipliers from L_p into L_q are found. We study the interaction of semigroups of the changing of signs of Fourier-Haar coefficients and that of functions.

The Haar system is not an unconditional basis of L_1. Therefore subspaces generated by a subsequence of χ in L_1 can be sufficiently complicated. On the other hand J. Gamlen - R. Gaudet characterized such subspaces in L_p $(1 < p < \infty)$. These results are contained in Chapter 15.

The Franklin system is closely connected with χ. We prove the criterion of equivalence of the Haar and the Franklin system in rearrangement invariant spaces in Chapter 16. In fact, we present T. Fiegel proof of Z. Cieselski's theorem about the equivalence of the Haar and the Franklin system in L_p $(1 < p < \infty)$.

Using χ, A. Olevskii constructed an orthogonal system with some interesting properties. It is studied in Chapter 17.

The bibliography devoted to the Haar system is very wide. It contains, in particular, the reference articles [112], [330], [339]. As far as we know there is no monograph completely devoted to the Haar system, though the monographs [40], [113], [132], [163], [288] contain separate chapters devoted to the Haar system. References given in our monograph are not complete. However, it contains not only papers quoted in the text but also many papers dealing with the Haar system.

Acknowledgements

This book was written under the support of the Russian Fund of Fundamental Investigations (grant 93-011-159).

We would like to express our gratitude to our colleagues B.I. Golubov, V.F. Gaposhkin, B.S. Kashin, E.M. Nikishin, A.M. Olevskii, K.I. Oskolkov, P.L. Ul'yanov, S.G. Krein, M.Z. Berkolaiko, V.I. Ovchinnikov, V.A. Rodin for their kind help, support and friendly criticism.

We are also greatful to Ya.A. Izrailevich, V.A. Khodurdkaya, V.N. Livina, E.S. Ukusova for their enthusiastic help during the preparation of the manuscript.

Remarks

Chapter 1.

A detailed account of measure theory is presented, for example, in [201]. The basic properties of a conditional expectation are proved in [202]. All statements of Section 1.b are borrowed from [162].

The notion of the decreasing rearrangement of any function appeared rather long ago (see, for example, [118]). At the same time it was observed that the decreasing rearrangements possessed extremal properties ((1.c.3) is the simplest of them). A development of theory of r.i. spaces stimulated investigations of such properties [154], [170].

All facts about r.i. spaces contained in Section 1.d are borrowed from the monographs [154], [163], [21]. The geometrical aspects (in particular, the notions of p-convexity and of q-concavity) are presented in [163]. One can read about Theorem 1.d.1 in [170]. Theorem 1.d.2 was proved by G.G. Lorentz and T. Shimogaki [166], [262]. One can read in [163] about Theorems 1.d.3, 1.d.4 and about stronger results in this direction. The statement of Remark 1.d.5 is obtained in [88].

Monographs [21], [23], [35], [154], [311] are devoted to the interpolation of linear (and non-linear) operators in Banach spaces and applications. The real and complex methods of interpolation are presented there. Theorems 1.e.1 and 1.e.2 are the classical ones of J. Marcinkiewicz and of M. Riesz. Theorem 1.e.3 is proved by A. Calderon and by B. Mityagin [52], [186]. Theorem 1.e.4 is obtained by D. Boyd [32], he also introduced the notion of indeces, which are now called the Boyd indeces. Theorems 1.e.5 and 1.e.6 go back to the classical results of G.H. Hardy, J.E. Littlewood, G. Polya on the doubly stochastic matrices, they are also connected with the paper of J.V.Ryff [241] about the doubly stochastic transformations (see [154], [23] for details).

Chapter 2.

The definition and the formula for partial sums of Fourier-Haar series were introduced by A. Haar in 1909 [117]. The estimates of the Fourier-Haar coefficients for $f \in C(0,1)$ were obtained by Z. Ciesielski [63]; the estimates for $f \in L_p(0,1)$, $1 \le p < \infty$ were proved by P.L. Ul'yanov [322].

Chapter 3.

The uniform convergence of Fourier-Haar series of continuous functions was proved by A. Haar [117]. The estimate of convergence rate by means of the continuity modulus was obtained by B. Szokefalvi-Nagy [294]. Later the right part of inequality (3.1) was revised by P.L. Ul'yanov [322] and by B.I. Golubov [109]. The left part of (3.1) is proved in [109]. Theorem 3.2 was obtained by P.L. Ul'yanov [322] and by Z. Ciesielski [63]. Theorems 3.4 and 3.6 were proved by P.L. Ul'yanov [322]; Theorem 3.5 was obtained by B.I. Golubov [109].

The linear spans of the first 2^n functions of the Haar and the Walsh systems coincide for every $n \in \mathbf{N}$. That is why the partial sums of order 2^n coincide also. The estimates of the approximation rate with respect to the Haar and the Walsh systems are presented in Monographs [113], [132]. The other results about approximation by the polynomials with respect to the H.s. can be found in [113], [292], [175], [347], [122], [288]. A lot of papers are devoted to the questions of convergence of the Haar series (see Reviews [112], [330], [339]).

Chapter 4.

The basis properties of the H.s. are investigated by J. Schauder [245] in L_p, $1 \le p < \infty$, by W. Orlicz [215] in Orlicz spaces, by H.W.Ellis, I.Halperin [87] in separable r.i. spaces. Theorem 4.a.1 has been proved by M.Z. Solomyak [289].

Theorem 4.a.4 is a particular case of the result of D.L. Doob (see [77], Theorem 3.2). Inequalities (4.a.3) for L_p, $p > 1$ are obtained by J. Marcinkiewicz [172]. For r.i. spaces inequalities (4.a.3) are proved independantly in [10] and in [207]. Lemma 4.a.7 for L_p, $p > 1$ is presented in [77] (see also [48]), the general case is investigated in [207].

Inequality 4.b.1 is proved by S.B. Stechkin [86]. Theorem 4.b.1 is obtained by V.G.Krotov [158]. The criterion of unconditionality of the H.s. in $L_p(w)$ is also proved in [158].

Chapter 5.

Theorem 5.1 is proved in [346]. Using the simplification of the method of R. Gundy [115], V.F. Guposhkin proved inequality (5.2) with 4 as a constant [102]. Inequality (5.2) with 3 as a constant is obtained in [342] (see also [125]). We follow the proof of D.L. Burkholder [46], [47].

Remark 5.3 was proved by L. Dor (see [45]). Inequality (5.7) with the exact constant is obtained in [43], the proof is borrowed from [46]. It is shown in [44] that inequality (5.7) is valid for the Haar series with coefficients belonging to an arbitrary Hilbert space.

Theorem. Let $1 < p < \infty$, and X is a Hilbert space. If a_k, $b_k \in X$ and $\|b_k\|_X \le \|a_k\|_X$ for $k \in \mathbf{N}$, then for every $n \in \mathbf{N}$

$$\Big(\int_0^1 \|\sum_{k=1}^n b_k \chi_k(t)\|_X^p \, dt\Big)^{1/p} \le (p^* - 1)\Big(\int_0^1 \|\sum_{k=1}^n a_k \chi_k(t)\|_X^p \, dt\Big)^{1/p}.$$

The unconditionality of the H.s. in L_p for $p \in (1, \infty)$ was obtained by J. Marcienkiewicz [171], [172] as a corollary of the result of R. Paley about the Walsh system [223].

Chapter 6.

Inequality (6.2) was presented for the first time in [346]. The proof is borrowed from [22]. About Theorem 6.3 see Remark to Theorem 8.8. Theorem 6.6 is proved for the Orlicz spaces in [99-101], for r.i. spaces it is presented in [154] and in [163].

Chapter 7.

Theorems 7.a.1 and 7.a.5 are similar to those for trigonometric series (see [358], Chapter 12). Theorem 7.a.8 is proved in [209]; Lemma 7.a.9, Corollaries 7.a.10 and 7.b.2 are obtained in [27]. Theorem 7.b.3 was proved by P.L. Ul'yanov [322]. Theorem 7.c.1 was obtained in a somewhat different form by B.I.Golubov [109]; as for Theorem 7.c.2 it is proved by S.V.Bochkarev in [27]. Theorem 7.c.3 is presented in [109]. The other estimates of Fourier-Haar coefficients of absolutely continuous functions are obtained in the paper by M.G.Robakidze [239]. The Fourier-Haar coefficients of the function with prescribed smoothness are investigated in [63], [176], [27]. Z.Ciesielski-J.Musielak [69], P.L.Ul'yanov [322], and B.I.Golubov [110] have found sufficient conditions of the convergence of the series $\sum_{n=1}^{\infty} |c_n(x)|^\alpha n^\beta$ for some values of α and β. The estimates of $c_n(x)$ for the functions of bounded variation were obtained by P.L. Ul'yanov [322], by L.G. Homutenko [120], by A.V. Maslov [174], by V.L. Matveev [176]. Functions with regular behavior of Fourier-Haar coefficients are investigated in [109], [156], [315], [209]. Reviews [112], [330], [339] contain more detailed information about the papers devoted to the Fouier-Haar coefficients.

Chapter 8.

Theorem 8.3 is due to B.Maurey [178]. Remark 8.5 is presented in [43] (see also [47]). Theorem 8.6 is proved independently in [10] and in [207]. Lemma 8.7 is due to D.L. Doob [77] . Theorem 8.8 for L_1 is proved in [72] (see also [50]); the generalization on Orlicz spaces is investigated in [49] (see also [48]; the case of an arbitrary r.i. space is considered in [128], [10], [206], [207]. Lemma 8.9 is obtained in [202] (see also [103]). Lemma 8.10 for the Orlicz spaces is proved in [202] and in [103], the general case of r.i. spaces is considered in [128] and in [207].

Chapter 9.

Theorem 9.2 was proved by A.M.Olevskii [214]. Lemma 9.4 was obtained in [240]. Theorem 9.7 for L_p, $1 < p < \infty$ was proved in [246], the case of an arbitrary r.i. space is presented in [163]. Remark 9.9 is borrowed from [163, p. 162].

Chapter 10.

Section 10.a is based on the results of paper [33]. The more general variants of Theorem 10.a.21 can be found in [34]. Theorem 10.a.22 is due to T. Ando [6]. Corollary 10.a.23 is presented in [80]. Theorem 10.a.24 was proved in [82].

In Section 10.b we follow [80]. In connection with Corollary 10.a.14 let us observe that in an arbitrary infinite dimensional Banach space there exists a unconditionable set of mutually non-equivalent bases [226]. Theorem 10.b.21 is proved in [8].

Chapter 11.

The rearrangements of the H.s. were investigated in many papers. V.F. Gaposhkin has proved that there exists a rearrangement $\pi \in \Pi$ such that the operator T_π is unbounded in L_p for $p \neq 2$ [98]. Ch. McCarthi has constructed the rearrangement $\pi \in \Pi$ such that T_π is an isometry in L_2 and $\|Q_e T_\pi Q_f\|_{L_2} = 1$ for every e, $f \subset [0,1]$ of positive measure, where Q_e is an operator of multiplication by the characteristic function of a set e. This result was extended to L_p for $p \in (1,\infty)$ in paper [256]; the case of $L_{p,q}$ for some values of p and q was considered in [254]. Theorem 11.5 is presented in [247], Theorem 11.6 was proved in [255], Theorem 11.10 was obtained in [248]. Theorem 11.17 was proved independantly by F. Shipp [264] and by P. Muller [195].

The other approaches to investigation of the operators rearranging the H.s. were developed in papers [90], [91]. The criterion of the boundedness of operators T_π in L_1 was found in [143]. Z. Ciesielski and S. Kwapien have proved that the shift operator $\chi_n \to \chi_{n+1}$ is bounded in L_p for $p \in (1,\infty)$ and is unbounded in L_1 [68].

Chapter 12.

Theorem 12.1 is a Corollary of the Theorem of S. Yano [346]. Theorem 12.2 is announced in [244]. Multipliers with respect to the H.s. were investigated by V.G. Krotov [157] and by O.B. Sadovskaya [242], [243].

Chapter 13.

The estimates of $(Ax)^*(t)$ for operators in pairs of L_p-type spaces appeared for the first time in the interpolation theorem of J. Marcinkiewicz [357]. They were reformulated in operator terms by A. Calderon [52]. The maximal Calderon operator was used in [154] for obtaining the optimal interpolation theorems. Inequality (13.1) is a particular case of that proved in [127] and in [250]. Theorems 13.1, 13.5, 13.6 are presented in [216].

Chapter 14.

The space wL appeared in the implicit form rather long ago. The classical Kolmogorov theorem states that the conjugate function operator maps L_1 in wL. After the appearence of the paper by A. Zygmund [357] about Marcinkiewicz interpolation theorem, such operators are called operators of weak type (1,1). Lemma 14.3 is well known.

Chapter 15.

Theorem 15.1 for L_p with $p \in (1, \infty)$ is proved in [97], its extention to r.i. spaces is presented in [163]. The rest results of this section are obtained in [204], [205]. The subsequences of the H.s. in the space $H^1(\delta) := \{f \in L_1 : \|S^\# f\|_{L_1} < \infty\}$ are investigated in [193]. Similar results for the finite dimensional case are obtained in [192]. The extension of the results from [97] and from [193] to the three-valued martingale difference sequence was proved in [194].

Chapter 16.

The Franklin system was introduced in [94]. Inequalities (16.a.1-2) and (16.e.7-8) are proved in [64, 65]. The equivalence of the Haar and the Franklin systems in L_p with $p \in (1, \infty)$ was estableshed in [70]. Our proof is borrowed from [91]. Let us observe also that in [91] it is proved that the H.s. is equivalent in the space L_p of vector-valued functions not only to the Franklin system but also to spline bases [67] and to wavelet systems [194]. Theorem 16.e.1 for L_1 was obtained in [268]. The necessity conditions of equivalence of the Haar and the Franklin systems in terms of dilation indices of the fundamental function of r.i. space were found in [144-147]. However, there exist r.i. spaces for which the dilation indices of the fundamental function are not equal to the Boyd indices. The final answer is obtained in [208].

Chapter 17.

The system $\{g_n\}$ has appeared in the paper by A.M. Olevskii [212]. K.S. Kazaryan showed that the system $\{g_n\}$ is not an unconditional basis of L_p with $p \neq 2$. Using the Franklin system, R. Zink [349], and independently K.S.Kazarian and S.A.Sargsian [139] have constracted the system of continuous functions with similar properties. The Olevskii system was used in many papers (see, for example, [230]).

Chapter 1

Preliminaries

1.a. Measure space

Let (S, F, μ) be a measure space. Here S is a set, F is a σ-algebra, μ is a σ-finite measure on F. If F_0 is a σ-subalgebra of F, $x \in L_1(S, F, \mu)$, then denote by E_{F_0} the unique, up to equivalence, F_0-measurable function satisfying

$$\int_A x\, d\mu = \int_A E_{F_0} x\, d\mu$$

for each $A \in F_0$. By the Radon – Nikodym theorem, such function exists. The function $E_{F_0}x = E_{F_0,\mu}x$ is called the conditional expectation with respect to F_0.

The operator E_{F_0} is positive, linear and has norm 1 in $L_1(S, F, \mu)$ and $L_\infty(S, F, \mu)$. For every $x \in L_\infty(S, F, \mu)$, $y \in L_1(S, F, \mu)$

$$E_{F_0}(xE_{F_0}y) = E_{F_0}xE_{F_0}y.$$

Therefore E_{F_0} is a projection. It is also sometimes called an averaging operator. Later on Σ denotes a σ-algebra of measurable subsets of [0,1], *mes e* or *mes*(e) is the Lebesque measure of e.

1.b. Main results on bases in Banach spaces

Let X be an infinite dimensional Banach space and $x_n \in X$, $n \geq 1$. A sequence $\{x_n\}$ is named a basis if, for each x , there is a unique sequence $\{a_n\}_1^\infty$, such that

$$x = \sum_{n=1}^\infty a_n x_n.$$

The order of elements $\{x_n\}_1^\infty$ is essential, generally speaking, in this definition.

A sequence $\{x_n\}_1^\infty$ is said to be a basic sequence if $\{x_n\}$ is a basis in $[x_n]_1^\infty$. Here $[x_n]_1^\infty$ denotes the closure in X of the linear span of $\{x_n\}_1^\infty$.

Each basis $\{x_n\}$ generates a sequence of projections P_n by the following equalities :

$$P_n(\sum_{i=1}^{\infty} a_i x_1) = \sum_{i=1}^{n} a_i x_i.$$

The projections P_n are called the natural projections associated to $\{x_n\}_1^\infty$. The open mapping theorem states that

$$\sup_n \|P_n\|_X < \infty.$$

This positive number is named a basis constant of $\{x\}_1^\infty$ and is denoted by $C_X(\{x_n\}_1^\infty)$. If $C_X(\{x_n\}_1^\infty) = 1$, then $\{x_n\}$ is said to be a monotone basis.

Theorem 1.b.1. *A sequence $\{x_n\}_1^\infty$ in a Banach space X is a basis iff*

1) $x_n \neq 0 \quad \forall n = 1, 2, \ldots$

2) There exists a constant $C > 0$ such that for every sequense $\{a_n\}_1^\infty$, and for every $1 \leq n < m < \infty$

$$\|\sum_{i=1}^{n} a_i x_i\|_X \leq C\|\sum_{i=1}^{m} a_i x_i\|_X;$$

3) The span $\{x_i\}$ is dense in X.

Clearly, a sequence $\{x_n\}$ is a basic one iff conditions 1-2 are satisfied. The smallest constant C in condition 2 coincides with $C_X(\{x_n\})$.

A basis $\{x_n\}$ is said to be normed if $\|x_n\|_X = 1$ for each $n = 1, 2, \ldots$

Two bases $\{x_n\}$ in X and $\{y_n\}$ in Y are called equivalent if the series $\sum x_n$ converges in X iff the series $\sum y_n$ converges in Y.

The closed graph theorem states that bases $\{x_n\}$ and $\{y_n\}$ are equivalent iff there exists a constant $C > 0$ such that for every finite sequence $\{a_n\}$

$$\frac{1}{C}\|\sum a_n y_n\|_Y \leq \|\sum a_n x_n\|_X \leq C\|\sum a_n y_n\|_Y.$$

Bases are named C-equivalent in this case and isometrically equivalent if $C = 1$.

Bases in a Banach space possess a stability property.

Theorem 1.b.2. *Let $\{x_n\}_1^\infty$ be a normed basis in a Banach space X, $y_n \in X$, $n = 1, 2, \ldots$ If*

$$\sum_{n=1}^{\infty} \|x_n - y_n\|_X \leq (2C_X(\{x_n\}))^{-1},$$

then $\{y_n\}$ is a basis in X and $\{x_n\}$ and $\{y_n\}$ are equivalent.

An analogous statement is true for basic sequences.
Let $\{x_n\}_1^\infty$ be a basic sequence, $y_n \in X$, $n = 1, 2, \ldots$ and

$$\sum \|x_n - y_n\|_X \le (8C(\{x_n\}))^{-1}.$$

If $[x_n]$ is complemented in X , then $[y_n]$ is complemented in X too.

Let $\{x_n\}_1^\infty$ be a basic sequence in X, $\{p_j\}_1^\infty$ be a strictly increasing sequence of integers, $\{a_n\}$ be a sequence of reals. Then the sequence

$$u_j = \sum_{n=p_j}^{p_{j+1}-1} a_n x_n$$

is called block-basis of $\{x_n\}$. Clearly, $\{u_j\}_1^\infty$ is a basic sequence too.

Theorem 1.b.3. *Let $\{x_n\}$ be a normed basis in a Banach space X and $\{x_n\}$ tends to 0 in X weakly. If X is isomorphic to a subspace of a Banach space Y with basis $\{y_n\}_1^\infty$, then there exists a subsequence $\{x_{n_j}\}$, which is equivalent to some block-basis of $\{y_n\}_1^\infty$.*

A series $\sum x_n$ is said to be unconditionally convergent in a Banach space X if, for each rearrangement π of integers, series $\sum x_{\pi(n)}$ converges in X.

Theorem 1.b.4. *A series $\sum x_n$ is unconditionally convergent iff $\sum \varepsilon_n x_n$ converges for all sequences $\varepsilon_n = \pm 1$.*

A basis $\{x_n\}$ in a Banach space X is said to be unconditional if, for any rearrangement π, $x_{\pi(n)}$ is a basis in X.

Theorem 1.b.5. *A basis $\{x_n\}$ in a Banach space X is unconditional iff there exists a constant $C > 0$ such that, for every integer N, for every set of signs $\{\varepsilon_n\}_1^N$, for every set of reals $\{a_n\}_1^N$ the following inequality holds*

$$\|\sum_{n=1}^N \varepsilon_n a_n x_n\|_X \le C\|\sum_{n=1}^N a_n x_n\|_X.$$

Consequently, each sequence $\{\varepsilon_n = \pm 1\}$ generates a bounded operator

$$T_\varepsilon(\sum a_n x_n) = \sum \varepsilon_n a_n x_n$$

and

$$UC_X(\{x_n\}) := \sup_{\varepsilon = \pm 1} \|T_\varepsilon\|_X < \infty.$$

The number $UC_X(\{x_n\})$ is named an unconditional basis constant. Clearly,

$$C_X(\{x_n\}) \le UC_X(\{x_n\}).$$

Let J be a subset of integers. Define the operator

$$P_J(\sum a_n x_n) = \sum_{n\in J} a_n x_n.$$

It is easy to show that

$$\sup_{J\subset \mathbf{N}} \|P_J\|_X \le \sup_{\varepsilon=\pm 1} \|T_\varepsilon\|_X \le 2 \sup_{J\subset \mathbf{N}} \|P_J\|_X.$$

Theorem 1.b.6. *Let $\{x_n\}_1^\infty$ be an unconditional basis in a Banach space X, $\{\lambda_n\}_1^\infty$ be a bounded sequence of reals and $\sum a_n x_n$ converges in X. Then*

$$\|\sum \lambda_n a_n x_n\|_X \le UC_X(\{x_n\}) \sup_n |\lambda_n| \, \|\sum a_n x_n\|_X.$$

Corollary 1.b.7. *Let $\{u_j\}_1^\infty$ be a block-basis of an unconditional basic sequence $\{x_n\}_1^\infty$ in a Banach space X. Then $\{u_j\}_1^\infty$ is an unconditional basic sequence and*

$$UC_X(\{u_n\}) \le UC_X(\{x_n\}).$$

Theorem 1.b.8. *Let $\{x_n\}_1^\infty$ be an unconditional basis in a Banach lattice X. There exists a constant C such that, for every integer n, for every set of reals $\{a_i\}_1^n$,*

$$\frac{1}{C}\|(\sum_{i=1}^n |a_i x_i|^2)^{1/2}\|_X \le \|\sum_{i=1}^n a_i x_i\|_X \le C\|(\sum_{i=1}^n |a_i x_i|^2)^{1/2}\|_X.$$

Remark 1.b.9. *The assumption that $\{x_n\}_1^\infty$ is an unconditional basis in X can not be replaced by the assumption that $\{x_n\}$ is an unconditional basic sequence in X.*

To prove this it is sufficient to consider the Rademacher functions in $L_\infty[0,1]$. However, that kind of generalization of Theorem 1.b.8 is true if $[x_n]$ is complemented in X or X is q-concave for some $q < \infty$ (the definition of q-concavity is contained in Section 1.d).

1.c. Rearrangements of Functions

If $x(t)$ is a measurable function on $[0,1]$ with respect to Lebesque measure, then its distribution function η_x is given by

$$\eta_x(\tau) = mes\{t : x(t) > \tau\}.$$

Two functions x and y are said to be equimeasurable if they have the same distribution function, that is, if $\eta_x(\tau) = \eta_y(\tau)$ for all τ. The distribution function η_x is non-negative, decreasing, and right-continuous on $(-\infty, \infty)$.

The decreasing rearrangement of x is the function x^* defined on $[0,1]$ by

$$x^*(t) = inf\{\tau : \eta_{|x|}(\tau) \leq t\}.$$

The most important property of rearrangements is known as the Hardy-Littlewood inequality :

$$\left| \int_0^1 x(t)y(t)\,dt \right| \leq \int_0^1 x^*(t)y^*(t)\,dt. \qquad (1.c.1.)$$

Let $y(t) = \kappa_e(t)$ be a characteristic function of a measurable subset $e \subset [0,1]$. Then

$$\left| \int_e x(t)\,dt \right| \leq \int_0^{mes(e)} x^*(t)\,dt. \qquad (1.c.2.)$$

The last inequality may be written in a more exact form:

$$\sup_{mes(e)=\tau} \int_e |x(t)|\,dt = \int_0^\tau x^*(t)\,dt \qquad \forall \tau \in [0,1]. \qquad (1.c.3.)$$

If $x, y \in L_1$, $x(t)$, $y(t)$, $a(t)$ are decreasing functions on $[0,1]$,

$$\int_0^\tau x(t)\,dt \leq \int_0^\tau y(t)\,dt \qquad \forall \tau \in [0,1],$$

$a \geq 0$, then

$$\int_0^1 x(t)a(t)\,dt \leq \int_0^1 y(t)a(t)\,dt.$$

Lemma 1.c.1. *Let x, y be equimeasurable functions on $[0,1]$, $\varepsilon > 0$. There exists a measure preserving mapping ω of $[0,1]$ into itself such that $\|x - y \circ \omega\|_{L_\infty} < \varepsilon$.*

1.d. Rearrangement Invariant Spaces

A Banach functional space E on $[0,1]$ with the Lebesque measure is said to be a rearrangement invariant (r.i.) space, if the following condition holds : if x, y are measurable functions, $y \in E$, $x^* \leq y^*$, then $x \in E$ and $\|x\|_E \leq \|y\|_E$.

In addition, we shall assume that an r.i. space E is separable or isomorphic to a separable space. In that case the norm of E is semicontinuous in the following sence. If $x \in E$, $0 \leq x_1 \leq x_2 \leq \dots \leq x$, $lim_{n\to\infty} x_n(t) = x(t)$ almost everywhere, then

$$\|x\|_E = lim_{n\to\infty} \|x_n\|_E.$$

Let us consider some examples of r.i. spaces.

1. L_p - spaces with the norm

$$\|x\|_{L_p} = \begin{cases} (\int_0^1 \mid x(t) \mid^p \, dt)^{\frac{1}{p}}, & 1 \leq p < \infty; \\ ess\sup_{0\leq t\leq 1} \mid x(t) \mid, & p = \infty. \end{cases}$$

2. Ozlicz spaces. Let $M(u)$ be an even convex function on the exis, $M(0) = 0$. Denote by L_M a space with the norm

$$\|x\|_{L_M} = \inf\{\lambda : \lambda > 0; \quad \int_0^1 M(\lambda^{-1}x(t))\, dt \leq 1\}.$$

3. Lorentz spaces. Let $\varphi(t)$ be an increasing concave function on $[0,1]$, $\varphi(0) = 0$. The norm in a Lorentz space $\Lambda(\varphi)$ is defined to be

$$\|x\|_{\Lambda(\varphi)} = \int_0^1 x^*(t)\, d\varphi(t).$$

Spaces with the norms

$$\|x\|_{L_{p,q}} = \left(\int_0^1 (x^*(t)t^{\frac{1}{p}})^q \frac{dt}{t}\right)^{\frac{1}{q}}, \quad 1 \leq p, q < \infty$$

with the usual modification for $q = \infty$ are named Lorentz spaces too.

The spaces $L_{p,q}(1 < p, q < \infty)$ are strictly normed. This statement fails for $q = 1$ or $q = \infty$.

The unit sphere of $L_{p,1}(1 < p < \infty)$ has a simple structure. The functions

$$(mes\, e_1 + mes\, e_2)^{-\frac{1}{p}}(\kappa_{e_1}(t) - \kappa_{e_2}(t)), \tag{1.d.1}$$

where e_1, e_2 are disjoint subsets of $[0,1]$, are extremal points of the unit sphere of $L_{p,1}$. The inverse statement is true. Each extremal point of the unit sphere of $L_{p,1}$ has the form (1.d.1).

Let A be a linear or quasi-linear operator acting from $L_{p,1}$ in a Banach space B and for some constant $C > 0$

$$\|A\kappa_e\|_B \leq C(mes\, e)^{\frac{1}{p}} \quad \forall e \in [0,1].$$

Then

$$\|Ax\|_B \leq 2C\|x\|_{L_{p,1}}$$

for every step function x. If, in addition, an operator A is positive, then

$$\|Ax\|_B \leq C\|x\|_{L_{p,1}}.$$

The norm of a positive function x in the spaces L_p and Lorentz ones may be expressed through its distribution function . We have

$$\|x\|_{L_p} = \left(p\int_0^\infty \eta_{|x|}(\tau)\tau^{p-1}\, d\tau\right)^{\frac{1}{p}}, \tag{1.d.2}$$

$$\|x\|_{\Lambda(\varphi)} = \int_0^\infty \varphi(\eta_{|x|}(\tau))\, d\tau. \qquad (1.d.3)$$

4. Marcinkiewich spaces. Let $\varphi \geq 0$, $\varphi(t)$ and $\frac{t}{\varphi(t)}$ are increasing functions on $(0,1]$. Such fuction is called a quasi-concave one. Given a quasi-concave function φ , there exists a concave function φ_1 such that $\varphi_1 \leq \varphi \leq 2\varphi_1$. Put

$$\|x\|_{M(\varphi)} = \sup_{e \subset [0,1],\quad mes\, e > 0} \frac{\int_e |\, x(t)\, |\, dt}{\varphi(mes\, e)} = \sup_{0<\tau\leq 1} \frac{\int_0^\tau x^*(t)\, dt}{\varphi(\tau)}.$$

The spaces $M(\varphi)$ are named Marcinkiewhich ones.

If $\varphi(t)$ is continuous in 0, then $\Lambda(\varphi)^* = M(\psi)$, where $\psi(t) = \frac{t}{\varphi(t)}$. Denote by $M_0(\psi)$ the closure of L_∞ in $M(\psi)$. If

$$\lim_{t\to 0} \frac{\psi(t)}{t} = \infty, \qquad (1.d.4)$$

then $M_0(\psi)$ is separable. It coinsides with the set of $x \in M(\psi)$ such that

$$\lim_{\tau\to 0} \frac{1}{\psi(t)} \int_0^\tau x^*(t)\, dt = 0.$$

Under condition (1.d.4) $(M_0(\psi))^* = \Lambda(\varphi)$.

Given an r.i. space E, we put

$$E' = \{x : x \in L_1, \quad \int_0^1 x(t)y(t)\, dt < \infty \quad \forall y \in E\}$$

and

$$\|x\|_{E'} = \sup_{\|y\|_E \leq 1} \int_0^1 x(t)y(t)\, dt.$$

The Banach space E' is called the aassociate space of E.

Denote by E_0 the closure of L_∞ in E.

Theorem 1.d.1. *Let E be a separable r.i. space. Every $y \in E'$ generates the linear functional on E :*

$$\langle x, y\rangle = \int_0^1 x(t)y(t)\, dt, \quad x \in E.$$

This mapping is an isometric isomorphism of E' into E^*.

Consequently, under the assumption $E \neq L_\infty$

$$(E_0)' = (E_0)^* = E'.$$

It is evident that $L'_\infty = L_1$.

For every $p \in (1,\infty), q \in [1,\infty]$

$$L'_{p,q} = L_{p',q'},$$

where, as usually, $\frac{1}{p} + \frac{1}{p'} = 1$.

If E is an r.i. space, then $\|\kappa_e\|_E$ depends only on the measure of $e \subset [0,1]$. It means that there exists a function φ_E named the fundamental function of E such that

$$\|\kappa_e\|_E = \varphi_E(mes\, e) \quad \forall e \in [0,1].$$

The fundamental function of any r.i. space E is quasi-concave and, consequently, is concave up to factor 2.

Suppose that E is an r.i. space and φ_E is concave, $\varphi(1) = 1$. Then

$$\Lambda(\varphi_E) \subset E \subset M(\psi_E),$$

where $\psi_E(t) = \frac{t}{\varphi_E(t)}$. The imbedding operator norm is equal to 1.

Denote

$$\sigma_\tau x(t) = \begin{cases} x(\tau^{-1}t), \quad 0 \le t \le 1; & \tau \ge 1; \\ \begin{cases} x(\tau^{-1}t), & 0 \le t \le \tau; \\ 0, & \tau < t \le 1; \end{cases} & 0 < \tau \le 1; \end{cases}$$

Given a number $\tau > 0$ and r.i. space E, the operator σ_τ acts in E and

$$\min(1,\tau) \le \|\sigma\|_E \le \max(1,\tau).$$

Since the function $\|\sigma\|_E$ is semi-multiplicative, then

$$\lim_{\tau\to 0} \frac{\ln\|\sigma_\tau\|_E}{\ln\tau} \quad \text{and} \quad \lim_{\tau\to\infty} \frac{\ln\|\sigma_\tau\|_E}{\ln\tau}$$

exist.

The numbers

$$\alpha_E = \lim_{\tau\to 0} \frac{\ln\|\sigma_\tau\|_E}{\ln\tau}, \quad \beta_E = \lim_{\tau\to 0} \frac{\ln\|\sigma_\tau\|_E}{\ln\tau}$$

are called Boyd indeces of an r.i. space E.

The inequalities $0 \le \alpha_E \le \beta \le 1$ are valid.

For example, if $E = L_{p,q}$, then $\|\sigma_\tau\|_E = \tau^{\frac{1}{p}}$. Hence

$$\alpha_{L_{p,q}} = \beta_{L_{p,q}} = \frac{1}{p}$$

As

$$\|\sigma_\tau\|_{E'} = \tau\|\sigma_{\frac{1}{\tau}}\|_E \quad \forall\tau > 0,$$

then

$$\alpha_{E'} = 1 - \beta_E, \quad \beta_{E'} = 1 - \alpha_E.$$

Let $p \geq 1$ be a number. A Banach lattice E is named p-convex if there exists a number $C > 0$ such that for every $n = 1, 2, \ldots$ and every $x_1, x_2, \ldots, x_n \in E$

$$\left\| \left(\sum_{k=1}^{n} | x_k |^p \right)^{\frac{1}{p}} \right\|_E \leq C \left(\sum_{k=1}^{n} \|x_k\|_E^p \right)^{\frac{1}{p}}. \tag{1.d.5}$$

The constant of p-convexity of E is defined as $\inf C$ in the last inequality.

A Banach lattice is said to be q-concave if

$$\left(\sum_{k=1}^{n} \|x_k\|_E^q \right)^{\frac{1}{q}} \leq C \left\| \left(\sum_{k=1}^{n} | x_k |^q \right)^{\frac{1}{q}} \right\|_E \quad \forall n, x_1, \ldots, x_n \in E.$$

If there exists a Banach lattice F such that

$$\|x\|_E = \| \, | x |^p \, \|_F^{\frac{1}{p}},$$

then E is p-convex. It implies that $L_{r,q}$ is p-convex if $p < \min(r, q)$. In fact, $L_{r,\infty}$ is p-convex if $p < r$ and is not r-convex. An Orlicz space L_M is p-convex iff the function $M(u^{\frac{1}{p}})$ is convex up to equivalence.

A Banach lattice E is p-convex iff E^* is p'-concave, where $\frac{1}{p} + \frac{1}{p}' = 1$. If an r.i. space E is p-convex (q-concave), then $\beta_E \leq \frac{1}{p}$ $\quad (\alpha_E \geq \frac{1}{q})$.

Many important properties of r.i. spaces are discribed with the help of the Boyd indeces. For example , we have :

Theorem 1.d.2. *The Hardy operator*

$$Hx(t) = t^{-1} \int_0^t x(s)\, ds$$

is bounded in an r.i. space E iff $\beta_E < 1$.

One can show that

$$\|H\|_E \leq \int_0^1 \|\sigma_{1/\tau}\|_E \, d\tau. \tag{1.d.6}$$

If $\beta_E < 1$, then the last integral converges . By the duality arguments , the condition $\alpha_E > 0$ is equivalent to the boundedness of the operator

$$H_1 x(t) = \int_t^1 s^{-1} x(s)\, ds$$

in an r.i. space E. Consequently, the operator

$$H_2 x(t) = \int_0^1 \frac{x(s)}{t + s}\, ds$$

is bounded in an r.i. space E iff $0 < \alpha_E \leq \beta_E < 1$.

Let $0 < \gamma < 1$ and $\alpha_E = \beta_E = \gamma$. Then for each $\varepsilon > 0$ and for every sufficiently small $\tau > 0$

$$\tau^{\gamma+\epsilon} \leq \|\sigma_\tau\|_E \leq \tau^{\gamma-\varepsilon}.$$

The assumption $\alpha_E = 0$ $(\beta_E = 1)$ shows that E and L_∞ (L_1) are similar to each other in some sense.

Theorem 1.d.3. *Let E be an r.i. space, $\alpha_E = 0$ $(\beta_E = 1)$, N be an integer. There exists a set of disjoint equimeasurable functions $\{x_i\}$, $\|x_i\|_E = 1$, $1 \leq i \leq N$ such that $\{x_i\}$ is equivalent to the unit vector basis of l_∞^N (l_1^N).*

The inverse statement is true . We mention that $\alpha_E > 0$ iff $\lim_{\tau\to 0} \|\sigma_\tau\|_E = 0$, $\beta_E < 1$ iff $\lim_{\tau\to\infty} \frac{1}{\tau}\|\sigma_\tau\|_E = 0$.

For every r.i. space E we have $L_\infty \subset E \subset L_1$ and

$$\|x\|_{L_1} \leq \|x\|_E \leq \|x\|_{L_\infty} \quad \forall x \in L_\infty. \tag{1.d.7}$$

Theorem 1.d.4. *Let E be an r.i. space and $0 < \alpha_E \leq \beta_E < 1$, let X be a complemented subspace of E. If X contains a complemented subspace isomorphic to E, then X is isomorphic to E.*

Remark 1.d.5. *The statement of Theorem 1.d.4. is valid for L_1.*

1.e. Interporlation Methods

Let $\bar{A} = (A_0, A_1)$ be a pair of Banach spaces continuously imbedding in some Hausdorff topological vector space $\mathcal{A}$. The set of such $x \in \mathcal{A}$ that are representable in the form $x = x_0 + x_1$ for some $x_0 \in A_0$, $x_1 \in A_1$ is called the sum of A_0 and A_1 and is denoted by $A_0 + A_1$. For $x \in A_0 + A_1$, set

$$\|x\|_{A_0+A_1} = \inf(\|x_1\|_{A_0} + \|x_1\|_{A_1}), \tag{1.e.1}$$

where the infimum is taken over all representations $x = x_0 + x_1$, $x_0 \in A_0$, $x_1 \in A_1$. For each $x \in A_0 \cap A_1$, set

$$\|x\|_{A_0\cap A_1} = \max(\|x\|_{A_0}, \|x\|_{A_1}).$$

Let $\bar{A} = (A_0, A_1)$, $\bar{B} = (B_0, B_1)$, (A, B) be three Banach pairs. Suppose that

$$A_0 \cap A_1 \subset A \subset A_0 + A_1,$$
$$B_0 \cap B_1 \subset B \subset B_0 + B_1.$$

A pair (A,B) is named an interporlation one with respect to $\bar{A},\bar{B}$ if each linear bounded operator T from A_i into B_i $(i=0,1)$ is bounded from A in B and

$$\|T\|_{A,B} \le C \max_{i=0,1} \|T\|_{A_i,B_i},$$

where C does not depend on T. We shall write $(A,B) \in int(\bar{A},\bar{B})$ in this case. If $A_0 = B_0$, $A_1 = B_1$ and $A = B$, we will say briefly that A is an interpolation or C-interpolation space with respect to (A_0,A_1).

Let (A_0,A_1) be a compatible pair. The K-functional is defined for each $x \in A_0 + A_1$ and $t > 0$ by

$$K(t,x,\bar{A}) = K(t,x) = \inf\left(\|x_0\|_{A_0} + t\|x_1\|_{A_1}\right),$$

where the infimum is taken over all representations of $x = x_0 + x_1$, $x_0 \in A_0$, $x_1 \in A_1$. It is evident that $K(t,x,\bar{A})$ is equivalent to $\|x\|_{A_0+A_1}$.

Given $\Theta \in (0,1), p \in [1,\infty]$, denote by $(A_0,A_1)_{\Theta,p}$ the set $x \in A_0 + A_1$, for which the functional

$$\|x\|_{\Theta,p} = \begin{cases} \left(\int_0^\infty (K(t,x)t^{-\Theta})^p \frac{dt}{t}\right)^{\frac{1}{p}}, & p < \infty; \\ \sup_{t>0} K(t,x)t^{-\Theta}, & p = \infty; \end{cases}$$

is finite. A pair $((A_0,A_1)_{\Theta,p},(B_0,B_1)_{\Theta,p})$ is an interpolation one with respect to $(\bar{A},\bar{B})$. More exactly, if $T \in L(A_0,B_0) \cup L(A_1,B_1)$, $\Theta \in (0,1)$, $1 \le p \le \infty$, then

$$\|T\|_{(A_0,A_1)_{\Theta,p},(B_0,B_1)_{\Theta,p}} \le \|T\|^{1-\Theta}_{A_0,B_0} \|T\|^{\Theta}_{A_1,B_1}.$$

The mapping $(0,1) \times [1,\infty) \times \bar{A} \to (A_0,A_1)_{\Theta,p}$ is called the functor of real interpolation.

Let A_0, A_1 be r.i. spaces, $x \in A_0 + A_1$, $x \ge 0$. Then the infimum in (1.e.1.) may be taken over non-negative $x_0 \in A_0$, $x_1 \in A_1$.

The real method of interpolation leads to

Theorem 1.e.1. *Let T be a linear bounded operator from L_{p_i} into $L_{q_i,\infty}$ and $p_i \le q_i$ $(i=0,1)$. Denote*

$$\frac{1}{p_\Theta} = \frac{(1-\Theta)}{p_0} + \frac{\Theta}{p_1}, \quad \frac{1}{q_\theta} = \frac{1-\Theta}{q_0} + \frac{\Theta}{q_1}, \quad 0 < \Theta < 1.$$

Then T is bounded from L_{p_Θ} into L_{q_Θ} and

$$\|T\|_{L_{p_\Theta},L_{q_\Theta}} \le C(\Theta)\|T\|^{1-\Theta}_{L_{p_0},L_{q_0}} \|T\|^{\Theta}_{L_{p_1},L_{q_1}}.$$

Now we present the complex method of interpolation. Let $\bar{A} = (A_0,A_1)$ be a compatible pair. We denote by $F(\bar{A})$ the space of functions $f(z)$ with values in $A_0 + A_1$, analytic in the strip $0 < Rez < 1$, and continuous and bounded in the closed strip, $f(i\tau) \in A_0$, $f(1+i\tau) \in A_1$ $\forall \tau \in (-\infty, +\infty)$, and equipped with the norm

$$\|f\|_{F(\bar{A})} = \max(\sup_\tau \|f(i\tau)\|_{A_0}, \quad \sup_\tau \|f(1+i\tau)\|_{A_1}).$$

We denote by $[A_0, A_1]_\Theta$ $(0 < \Theta < 1)$ the collection of all elements $x \in A_0 + A_1$ representable in the form $x = f(\Theta)$ for some $f \in F(\bar{A})$. In this set we introduce the norm

$$\|x\|_{[A_0,A_1]_\Theta} = \inf_{f(\alpha)=x} \|f\|_{F(\bar{A})}.$$

The mapping $(0,1) \times \bar{A} \to [A_0, A_1]_\Theta$ is named the functor of complex interporlation. If T is bounded from A_i in B_i $(i = 0, 1)$, then

$$\|T\|_{[A_0,A_1]_\Theta,[B_0,B_1]_\Theta} \leq \|T\|^{1-\Theta}_{A_0,B_0} \|T\|^{\Theta}_{A_1,B_1}.$$

From the above statement we have

Theorem 1.e.2. *Let T be a linear bounded operator from L_{p_i} into L_{q_i} $(i = 0, 1)$ and*

$$\frac{1}{p_\Theta} = \frac{1-\Theta}{p_0} + \frac{\Theta}{p_1}, \quad \frac{1}{q_\Theta} = \frac{1-\Theta}{q_0} + \frac{\Theta}{q_1}, \qquad 0 < \Theta < 1$$

Then T is bounded from L_{p_Θ} into L_{q_Θ} and

$$\|T\|_{L_{p_\Theta},L_{q_\Theta}} \leq 2\|T\|^{1-\Theta}_{L_{p_0},L_{q_0}} \|T\|^{\Theta}_{L_{p_1},L_{q_1}}.$$

If, in addition, $p_0 \leq q_0$, $p_1 \leq q_1$, then factor 2 may be omitted in the last inequality.

In general, the real and compex methods of interpolation do not give a full description of the set of interpolation spaces. However, the set of interpolation spaces with respect to the pair (L_1, L_∞) can be described. Since L_1 and L_∞ are r.i. spaces, then every interpolation space with the interpolation constant 1 is an r.i. one.

Theorem 1.e.3. *Let $C > 1$, $x, y \in L_1$, E be an r.i. space. Then E is a C-interpolation space between L_1 and L_∞ iff the inequalities*

$$\int_0^\tau x^*(t)\,dt \leq \int_0^\tau y^*(t)\,dt \quad \forall \tau \in [0,1] \tag{1.e.2}$$

imply

$$\|x\|_E \leq C\|y\|_E.$$

As we assume that an r.i. space is separable or conjugate to some separable one, then it satisfies this condition with $C = 1$.

Theorem 1.e.4. *Let E be an r.i. space, $1 < p_0 < p_1 < \infty$ and $\frac{1}{p_1} < \alpha_E \leq \beta_E < \frac{1}{p_0}$. If a linear operator A is bounded in L_{p_0} and L_{p_1} then A is bounded in E.*

We introduce a useful notation of semiordering. We shall write $x \prec y$ if (1.e.2.) is satisfied. The rearrangements properties imply

Theorem 1.e.5. *Let E be an r.i. space. If $y \in E$, $x \prec y$, then $x \in E$ and $\|x\|_E \leq \|y\|_E$.*

Clearly, $E_F x \prec x$ for each $x \in L_1$ and each σ-algebra F. Therefore we have

Theorem 1.e.6. *Let E be an r.i. space, F be a σ - algebra. The operator E_F is bounded in E and*

$$\|E_F\|_E = 1.$$

Chapter 2

Definition and Main Properties of the Haar System

Let Ω_1 be the set of (n,k) such that $1 \le k \le 2^n$, $n = 0,1,\ldots;$ $\Omega :=$ $\Omega_1 \cup (0,0)$. Denote

$$\Delta_0^0 = (0,1), \qquad \bar{\Delta}_0^0 = [0,1],$$

$$\Delta_n^k = \left(\tfrac{k-1}{2^n}, \tfrac{k}{2^n}\right), \qquad \bar{\Delta}_n^k = \left[\tfrac{k-1}{2^n}, \tfrac{k}{2^n}\right], \quad (n,k) \in \Omega_1.$$

Such intervals are named dyadic. Clearly, if two dyadic intervals intersect, then one of them contains the other. The inclusion $\Delta_n^k \supset \Delta_p^q$ is equivalent to conditions $p \ge n$, $2^{p-n}(k-1) < q \le 2^{p-n}k$.

Put $\chi_0^0(t) \equiv 1$. If $(n,k) \in \Omega_1$, then

$$\chi_n^k(t) = \begin{cases} 0, & t \in \bar{\Delta}_n^k; \\ 1, & t \in \Delta_{n+1}^{2k-1}; \\ -1, & t \in \Delta_{n+1}^{2k}. \end{cases}$$

The value of $\chi_n^k(t)$ in a discontinuity point t is defined as

$$\chi_n^k(t) = \lim_{\varepsilon \to 0} \frac{1}{2} \left(\chi_n^k(t-\varepsilon) + \chi_n^k(t+\varepsilon)\right).$$

If $k=1$ or $k=2^n$, then the value $\chi_n^k(t)$ in 0 and 1 is defined so that $\chi_n^k(t)$ is continuous in 0 and 1. The set of functions $\chi_n^k(t), (n,k) \in \Omega_1$ is called the Haar system (H.s.). Sometimes it is convenient to use the single-index H.s. The one-to-one mapping of Ω on the set of integers is realized by the following:

$$\begin{aligned} &(0,0) \to 1, \\ &(2^n, k) \to m = 2^n + k, \quad (n,k) \in \Omega_1. \end{aligned}$$

The collection $\{\chi_n^k\}_{k=1}^{2^n}$ is named an n-bundle. Clearly,

$$\int_0^1 \chi_n^k(t)\chi_p^q(t)\,dt = \begin{cases} 2^{-n}, & p = n, \quad q = k; \\ 0\ , & (p,q) \ne (n,k). \end{cases}$$

Therefore the system $\{2^{\frac{n}{2}}\chi_n^k(t), \quad (n,k) \in \Omega\}$ is orthonormal.

Let $m = 2^n + k, \quad (n,k) \in \Omega_1$. Denote by A_m the set of intervals :

$$A_1 = \{(0,1)\};$$

$$A_m = \{\Delta_{n+1}^1, \Delta_{n+1}^2, ..., \Delta_{n+1}^{2k}, \Delta_n^{k+1}, ..., \Delta_n^{2^n}\}. \tag{2.1}$$

Denote by $\mathcal{F}_m$ the σ-algebra generated by partitioning A_m. Clearly, each function $\{\chi_i\} \quad (1 \leq i \leq m)$ is measurable with respect to $\mathcal{F}_m$, and $\mathcal{F}_m$ is the smallest σ-algebra. Given $m \geq 1$, the polynomial

$$x(t) = \sum_{j=1}^{m} c_j \chi_j(t), \quad m \geq 1$$

with respect to the H.s. is a step function. It is a constant on each interval from A_m. It satisfies the condition

$$x(t) = \frac{1}{2} \lim_{\delta \to 0} (x(t-\delta) + x(t+\delta)) \quad \forall t \in (0,1),$$

and $x(t)$ is continuous in 0 and 1. Denote by D_m the set of such functions. Since $\cup_m D_m$ is dence in $L_p, \quad 1 \leq p < \infty$, then the Haar system is a complete one in L_p.

Let $E_{\mathcal{F}_m}$ be a conditional expactation with respect to A_m :

$$E_{\mathcal{F}_1} x(t) = \int_0^1 x(s)\, ds;$$

if $m = 2^n + k, \quad (n,k) \in \Omega$, then

$$\begin{aligned} E_{\mathcal{F}_m} x(t) &= \textstyle\sum_{i=1}^{2k} 2^{-n-1} (\int_{\Delta_{n+1}^i} x(s)\, ds)\ \kappa_{\Delta_{n+1}^i}(t) + \\ &+ \textstyle\sum_{j=k+1}^{2^n} 2^{-n} \int_{\Delta_n^j} x(s)\, ds\ \kappa_{\Delta_n^j}(t), \quad t \in [0,1]. \end{aligned}$$

It follows from the definitions of the H.s. and the dyadic intervals that

$$E_{\mathcal{F}_m}(\chi_{m+1}) = 0 \quad \forall\, m = 1, 2, ... \quad . \tag{2.2}$$

Given $x \in L_1$, the Fourier - Haar coefficients are defined by

$$c_1(x) = \textstyle\int_0^1 x(s)\, ds;$$

$$c_m(x) = c_{n,k}(x) = 2^n \textstyle\int_0^1 x(s) \chi_n^k(s)\, ds =$$

$$= 2^n \left(\textstyle\int_{\Delta_{n+1}^{2k-1}} x(s)\, ds - \int_{\Delta_{n+1}^{2k}} x(s)\, ds \right) =$$

$$= 2^m \textstyle\int_{\Delta_{n+1}^{2k-1}} (x(s) - x(s + 2^{-n-1}))\, ds,$$

where $m = 2^n + k, \quad (n,k) \in \Omega_1$.

Using standard arguments we can obtain the representation of the partial sum of Fourier - Haar series

$$S_m x(t) := \sum_{i=1}^{m} c_i(x)\chi_i(t) = \int_0^1 K_m(t,s)x(s)\,ds,$$

where

$$K_m(t,s) = \sum_{i=1}^{m} \frac{1}{\|\chi_i\|_{L_1}} \chi_i(t)\chi_i(s).$$

It is easy to check that, for $m = 2^n + k, \ (n,k) \in \Omega_1$,

$$K_m(t,s) = \begin{cases} 2^{n+1}, & (t,s) \in \Delta_{n+1}^i \times \Delta_{n+1}^i, \quad 1 \le i \le 2k; \\ 2^n, & (t,s) \in \Delta_n^i \times \Delta_n^i, \quad k < i \le 2^n; \\ 0, & \text{a.e. for other} \quad (t,s) \in [0,1] \times [0,1]. \end{cases}$$

Hence,

$$S_m x(t) = \begin{cases} 2^{n+1} \int_{\Delta_{n+1}^i} x(s)\,ds, & t \in \Delta_{n+1}^i, \ 1 \le i \le 2k; \\ 2^n \int_{\Delta_n^i} x(s)\,ds, & t \in \Delta_n^i, \ k < i \le 2^n, \end{cases}$$

where $m = 2^n + k, \quad (n,k) \in \Omega_1$. It means that

$$S_m x = E_{\mathcal{F}_m} x, \quad m = 1, 2, \ldots \ .$$

Let x be a summable function on $[0,1]$. We extend x to $(1,2]$ as a periodical function and put

$$\omega(\delta, x) \quad = \sup\nolimits_{0 \le t_1 < t_2 \le 1, \ \ t_1 - t_2 \le \delta} \mid x(t_1) - x(t_2) \mid, \quad 0 < \delta \le 1;$$

$$\omega_p(\delta, x) \quad = \sup\nolimits_{0 < \tau \le \delta} \left(\int_0^1 \mid x(t+\tau) - x(t) \mid^p dt \right)^{\frac{1}{p}}, \quad 0 < \delta \le 1.$$

The function $\omega(\delta,x) \quad (\omega_p(\delta,x))$ is named the module of continuity (module of continuity in L_p). It follows from (2.3) that for $(n,k) \in \Omega_1$

$$|c_{n,k}(x)| \le 2^n \int_{\Delta_{n+1}^{2k-1}} |x(s) - x(s + 2^{-n-1})|\,ds \le \frac{1}{2}\omega(2^{-n-1}, x);$$

$$\sum_{k=1}^{2^n} |c_{n,k}(x)| \le \sum_{k=1}^{2^n} \frac{1}{2}\omega(2^{-n-1}, x) = 2^{n-1}\omega(2^{-n-1}, x);$$

$$|\sum_{k=1}^{2^n} c_{n,k}(x)| = 2^n \sum_{k=1}^{2^n} \int_{\Delta_{n+1}^{2k-1}} (x(s) - x(s + 2^{-n-1})\, ds \leq 2^n \omega_1(2^{-n-1}, x).$$

Using the Holder inequality we can obtain the similar estimate for functions from L_p $(1 < p < \infty)$:

$$| c_{n,k}(x) | \leq 2^{\frac{n+1}{p}-1} \left(\int_{\Delta_{n+1}^{2k-1}} | x(s) - x(s + 2^{-n-1}) |^p \, dt \right)^{\frac{1}{p}}$$

and

$$(\sum_{k=1}^{2^n} |c_{n,k}(x)|^p)^{1/p} \leq$$

$$\leq 2^{\frac{n+1}{p}-1} (\sum_{k=1}^{2^n} \int_{\Delta_{n+1}^{2k-1}} |x(t) - x(t + 2^{-n-1})|^p \, dt)^{1/p} \leq$$

$$\leq 2^{\frac{n+1}{p}-1} \omega_p(2^{-n-1}, x),$$

where $(n, k) \in \Omega_1$.

Chapter 3

Convergence of Haar Series

One of the main propeties of the H.s. is that it forms a basis in C, L_p $(1 \leq p < \infty)$ and moreover in a separable r.i. space. Any function $\chi_n(t)$ $(n > 1)$ is discontinuous. Therefore if $x \in C[0,1]$, then the convergence $S_n x$ to x is meant in L_∞.

Theorem 3.1. *If $x \in C[0,1]$, m is an integer, then*

$$\frac{1}{6}\omega(\frac{1}{m}, x) \leq \|x - S_m x\|_{L_\infty} \leq 2\omega(\frac{1}{m}, x). \tag{3.1}$$

Proof. Let $m = 2^n + k$, $(n,k) \in \Omega_1$ and $t \in \Delta_{n+1}^i$, $1 \leq i \leq 2k$. By (2.4)

$$| x(t) - S_m x(t) |=| x(t) - 2^{n+1} \int_{\Delta_{n+1}^i} x(s)\, ds |\leq$$

$$\leq 2^{n+1} \int_{\Delta_{n+1}^i} | x(t) - x(s) | \; ds \leq \omega(2^{-n-1}, x).$$

Analogously, if $t \in \Delta_n^i$ $(k < i \leq 2^n)$, then

$$| x(t) - S_m x(t) |\leq \omega(2^{-n}, x).$$

Let $t_0 \in [0,1]$ be a dyadic number. The obtained inequalities imply

$$| x(t_0) - S_m x(t_0 + 0) |\leq \omega(2^{-n}, x);$$

$$| x(t_0) - S_m x(t_0 - 0) |\leq \omega(2^{-n}, x).$$

Since $S_m x(t_0)$ is situated between $S_m x(t_0 - 0)$ and $S_m x(t_0 + 0)$, then

$$| x(t_0) - S_m x(t_0) |\leq \omega(2^{-n}, x).$$

So, the inequality

$$| x(t) - S_m x(t) |\leq \omega(2^{-n}, x)$$

is established for every $t \in [0,1]$. Clearly,

$$\omega(2^{-n},x) \le \omega(\frac{2}{m},x) \le 2\omega(\frac{1}{m},x).$$

This proves the right part of (3.1).

To prove the left part of (3.1) denote the intervals of constancy of $S_m x(t)$ by $\delta_1, \delta_2, ..., \delta_m$. Let $\bar{\delta_1}, \bar{\delta_2}, ..., \bar{\delta_m}$ be the corresponding closed intervals, $m = 2^n + k$, $1 \le k \le 2^n$. We can find $0 \le t_1 < t_2 \le 1$ such that $| t_1 - t_2 | \le 2^{-n}$ and

$$| x(t_1) - x(t_2) | = \omega(2^{-n},x).$$

There exist integers $i, j \in \{1, 2, ..., m\}$, for which $t_1 \in \bar{\delta_i}$, $t_2 \in \bar{\delta_j}$. Then $i \le j \le i+2$. We may assume that $j = i+2$. The ends of $\bar{\delta}_{i+1}$ are denoted by t_3 and t_4. Since

$$| x(t_1) - x(t_2) | \le | x(t_1) - x(t_3) | + | x(t_3) - x(t_4) | + | x(t_4) - x(t_2) |,$$

we can chose an interval $\bar{\delta}_k$, $k \in \{i, i+1, i+2\}$ and some number $s_1, s_2 \in \bar{\delta}_k$ such that

$$| x(s_1) - x(s_2) | \ge \frac{1}{3}\omega(2^{-n},x).$$

Then

$$\|x - S_m x\|_{L_\infty} \ge \sup_{t \in \delta_k} | x(t) - S_m x(t) | \ge \frac{1}{6}\omega(2^{-n},x) \ge \frac{1}{6}\omega(\frac{1}{m},x).$$

□

The proof of Theorem 3.1 shows that

$$\frac{1}{4}\omega(2^{-n},x) \le \|x - S_{2^n} x\|_{L_\infty} \le \omega(2^{-n},x). \tag{3.2}$$

We note that the continuity of x was not used in the proof of the right part of (3.1). Therefore the right part of (3.1) is true for every $x \in L_\infty$. On the other hand, the continiuty of x is essential for the validity of the left side of (3.1). Indeed, let x be a Haar polynomial and its degree be bigger than 1. Then

$$\|x - S_m x\|_{L_\infty} = 0$$

for sufficiently big m and $\omega(\frac{1}{m},x) > 0$ for each m. An analogous statement for L_p is valid only partly.

Theorem 3.2. *Let $1 \le p < \infty$ and $x \in L_p$. Then*

$$\|x - S_m x\|_{L_p} \le (1+2^p)^{1/p} 2^{1/p} \omega_p(\frac{1}{m},x). \tag{3.3}$$

Proof. Applying (2.4) we have

$$\int_0^1 \mid x(t) - S_{2^n}x(t) \mid^p dt = \sum_{k=1}^{2^n} \int_{\Delta_n^k} \mid x(t) - 2^n \int_{\Delta_n^k} x(s)\,ds \mid^p dt \leq$$

$$\leq 2^n \sum_{k=1}^{2^k} \int_{\Delta_n^k} \int_{\Delta_n^k} \mid x(t) - x(s) \mid^p ds\,dt \leq$$

$$\leq 2^n \int_0^1 \int_{-2^n}^{2^n} \mid x(t) - x(t+u) \mid^p du\,dt =$$

$$= 2^n \int_{-2^{-n}}^{2^{-n}} \int_0^1 \mid x(t) - x(t+u) \mid^p dt\,du \leq$$

$$\leq 2^{n+1} \int_0^{2^{-n}} \omega_p^p(u,x)\,du \leq 2\omega_p^p(2^{-n},x).$$

Hence,

$$\|x - S_{2^n}x\|_{L_p} \leq 2^{\frac{1}{p}}\omega_p(2^{-n},x).$$

Repeating the arguments from Theorem 3.1, we obtaine the estimate

$$\|x - S_m x\|_{L_p} \leq (1+2^p)^{\frac{1}{p}} 2^{\frac{1}{p}} \omega_p(\frac{1}{m},x). \quad \square$$

Remark 3.3. Let E be a separable r.i. space. A function $x \in E$ defined on $[0,1]$ may be continued on (1,2] by periodicity. Denote $\tau_s x(t) = x(t+s)$, where $0 \leq t$, $s \leq 1$. The functional

$$\omega_E(\delta,x) = \sup_{0\leq s\leq\delta} \|\tau_s x - x\|_E, \quad 0 \leq \delta \leq 1$$

is said to be the continuity modulus of $x \in E$. The arguments from the proof of Theorem 3.2. leads to the analogous estimate

$$\|x - S_m x\|_E \leq 6\omega_E(\frac{1}{m},x), \quad m = 1,2,...$$

One can show that to estimate $\|x - S_m x\|_{L_p}$ below through $\omega_p(\frac{1}{m},x)$ is impossible, $1 \leq p < \infty$. However it is possible in different terms.

Theorem 3.4. *Let* $1 \leq p < \infty$, $x \in L_p$ *and*

$$x(t) = \sum_{(n,k)\in\Omega} a_{n,k}\chi_n^k(t)$$

Then

$$\left(\sum_{k=1}^{n} \mid a_{n,k} \mid^p\right)^{\frac{1}{p}} \leq 2^{\frac{n+p+1}{p}} \omega_p(2^{-n},x)$$

for each $n = 1,2,...$

Proof. Since

$$\sum_{k=1}^{2^n} a_{n,k}\chi_n^k = S_{2^{n+1}}x - S_{2^n}x,$$

then

$$\begin{aligned}(2^{-n}\textstyle\sum_{k=1}^{2^n}|a_{n,k}|^p)^{1/p} &= \|\textstyle\sum_{k=1}^{2^n} a_{n,k}\chi_n^k\|_{L_p} = \\ &= \|S_{2^{n+1}}x - S_{2^n}x\|_{L_p} \le \|x - S_{2^{n+1}}x\|_{L_p} + \|x - S_{2^n}x\|_{L_p}\end{aligned} \tag{3.4}$$

Using inequality (3.3) we get

$$(\sum_{k=1}^{2^n} | a_{n,k} |^p)^{\frac{1}{p}} \le 2^{\frac{n}{p}}(2^{\frac{1}{p}}\omega_p(2^{-n-1},x) + 2^{\frac{1}{p}}\omega_p(2^{-n},x)) \le 2^{\frac{n+p+1}{p}}\omega_p(2^{-n},x)$$

□

Theorem 3.5. *Let N be an integer, $1 \le p < \infty$, $x \in L_p$. Then*

$$\omega_p(2^{-N},x) \le 10 \cdot 2^{\frac{-N}{p}} \sum_{n=0}^{N+1} \|x - S_{2^n}x\|_{L_p}$$

Proof. Let

$$x(t) = \sum_{(n,k)\in\Omega} a_{n,k}\chi_n^k(t)$$

$$x_n(t) = \begin{cases} a_{0,0}\chi_0^0(t) + a_{0,1}\chi_0^1(t), & n = 0; \\ \sum_{k=1}^{2^n} a_{n,k}\chi_n^k(t), & n \ge 1. \end{cases}$$

As $\omega_p(\delta,x) \le 2\|x\|_{L_p}$ for every $\delta \in [0,1]$ then

$$\begin{aligned}\omega_p(2^{-N},x) &\le \omega_p(2^{-N},x - S_{2^N}x) + \omega_p(2^{-N},S_{2^N}x) \le \\ &\le 2\|x - S_{2^N}x\|_{L_p} + \textstyle\sum_{n=0}^{N}\omega_p(2^{-N},x_n)\end{aligned}$$

Clearly,

$$\omega_p(2^{-N},x_n) = \|x_n - \tau_{2^{-N}}x_n\|_{L_p}$$

for any $n = 0, 1, ...N$. Put $a_{n,2^n+1} = a_{n,1}$. We have

$$\begin{aligned}&\textstyle\sum_{k=1}^{2^n} a_{n,k}\chi_n^k(t) - \sum_{k=1}^{2^n} a_{n,k}\chi_n^k(t + 2^{-N}) = \\ &= 2\textstyle\sum_{k=1}^{2^n} a_{n,k}\kappa_{((2k-1)2^{-n-1}-2^{-N},(2k-1)2^{-n-1})}(t) - \\ &- \textstyle\sum_{k=1}^{2^n}(a_{n,k} + a_{n,k+1})\kappa_{k2^{-n}-2^{-N},k2^{-n})}(t),\end{aligned}$$

$$x_0(t) - x_0(t-2^{-N}) = 2a_{0,1}(\kappa_{(2^{-1}-2^{-N},2^{-1})}(t) - \kappa_{(1-2^{-N},1)}(t)).$$

Therefore

$$\|x_n - \tau_{2^{-N}}x_n\|_{L_p} = (\textstyle\sum_{k=1}^{2^n} |2a_{n,k}|^p 2^{-N} + \sum_{k=1}^{2^n} |a_{n,k} + a_{n,k+1}|^p 2^{-N})^{1/p} \leq$$

$$\leq 4 \cdot 2^{-\frac{N}{p}} (\textstyle\sum_{k=1}^{2^n} \mid a_{n,k} \mid^p)^{\frac{1}{p}}, \quad n = 1, 2, ..., N$$

and

$$\|x_0 - \tau_{2^{-N}}x_0\|_{L_p} = ((\mid 2a_{0,1} \mid^p + \mid 2a_{0,1} \mid^p)2^{-N})^{\frac{1}{p}} \leq 2^{2-\frac{N}{p}} \mid a_{0,1} \mid .$$

Using the obtained estimates and inequality (3.4), we get

$$\omega_p(2^{-N}, x) \leq 2\|x - S_{2^N}x\|_{L_p} + \textstyle\sum_{n=0}^{N} 2^{2-\frac{N}{p}} \left(\sum_{k=1}^{2^n} \mid a_{n,k} \mid^p\right)^{\frac{1}{p}} \leq$$

$$\leq 2\|x - S_{2^N}x\|_{L_p} + \textstyle\sum_{n=0}^{N} 2^{2-\frac{N}{p}+\frac{n}{p}} \left(\|x - S_{2^{n+1}}x\| + \|x - S_{2^n}x\|_{L_p}\right) \leq$$

$$\leq 10 \cdot 2^{-\frac{N}{p}} \textstyle\sum_{n=0}^{N+1} 2^{\frac{n}{p}} \|x - S_{2^n}x\|_{L_p}.$$

□

Let E be an r.i. space, $x \in E$ and let m be an integer. Put

$$F_m(x, E) = inf_{\{a_k\}} \|x - \sum_{k=1}^{m} a_k \chi_k\|_E.$$

Clearly,

$$F_m(x, E) \leq \|x - S_m x\|_E, \quad m = 1, 2, ...$$

This trivial estimate is exact up to a constant.

Theorem 3.6. *Let E be a separable r.i. space, $x \in E$ and let m be an integer. Then*

$$\|x - S_m x\|_E \leq 2F_m(x, E) \tag{3.5}$$

Proof. Recall that the set of intervals on which all functions $\chi_1, \chi_2, ..., \chi_m$ are constant is denoted by A_m (see (2.1)). By (2.4) and Theorem 1.e.6 $\|S_m\|_E = 1$. Therefore

$$\|I - S_m\|_E \leq 2 \tag{3.6}$$

Given a sequence of numbers $b_\delta, \quad \delta \in A_m$, consider the function

$$y(t) = x(t) - \sum_{\delta \in A_m} b_\delta \kappa_\delta(t)$$

Then

$$S_m y = S_m x - \sum_{\delta \in A_m} b_\delta \kappa_\delta(t)$$

and

$$y - S_m y = x - S_m x.$$

Using (3.6) we get

$$\|x - S_m x\|_E = \|y - S_m y\|_E \leq 2\|y\|_E = 2\|x - \sum_{\delta \in A_m} b_\delta \kappa_\delta\|_E.$$

Since $span\{\kappa_\delta,\ \delta \in A_m\} = span\{\chi_k,\ 1 \leq k \leq m\}$, then

$$\|x - S_m x\|_E \leq 2 \inf_{\{b_\delta\}} \|x - \textstyle\sum_{\delta \in A_m} b_\delta \kappa_\delta\|_E =$$

$$= 2 \inf_{\{a_k\}} \|x - \textstyle\sum_{k=1}^{m} a_k \chi_k\|_E = 2F_m(x, E).$$

□

Estimate (3.6) is precise on the set of r.i. spaces. If $E = L_1$, $x_n(t) = n\kappa_{(0,\frac{1}{n})}(t)$, $n = 1, 2, ...$ then $S_1 x_n = 1$ and

$$\lim_{n \to \infty} \|x_n - S_1 x_n\|_{L_1} = 2.$$

It means that $\|I - S_1\|_{L_1} = 2$. By duality argument $\|I - S_1\|_{L_\infty} = 2$. Clearly, $\|I - S_1\|_{L_2} = 1$. Applying the M.Riesz interpolation theorem (1.e.2), we get

$$\|I - S_1\|_{L_p} \leq 2^{\frac{2}{\min(p,p')} - 1}$$

where $1 < p < \infty$, $\frac{1}{p} + \frac{1}{p'} = 1$. Formula (3.6) may be written in an equuivalent form. Let $e_1, e_2, ..., e_m$ be disjoint subsets of $[0, 1]$, $a_1, a_2, ..., a_m$ be numbers, E be an r.i. space, $x \in E$ and

$$\int_{e_k} x(t)\, dt = 0 \quad \forall k = 1, 2, \ldots, m.$$

Then

$$\|x + \sum_{k=1}^{m} a_k \kappa_{e_k}\|_E \geq \frac{1}{2}\|x\|_E.$$

Chapter 4

Basis Properties of the Haar System

4.a. R.i. spaces

Theorem 3.2 shows that the H.s. forms a basis in L_p, $1 \le p < \infty$. This statement may be generalized.

Theorem 4.a.1. *Given an r.i. space E, an integer n and numbers $a_1, a_2, ..., a_{n+1}$, the following inequality holds*

$$\|\sum_{k=1}^{n+1} a_k \chi_k\|_E \ge \|\sum_{k=1}^{n} a_k \chi_k\|_E. \qquad (4.a.1)$$

Proof. The functions

$$\sum_{k=1}^{n} a_k \chi_k + a_{n+1}\chi_{n+1}, \quad \sum_{k=1}^{n} a_k \chi_k - a_{n+1}\chi_{n+1}$$

are equimeasurable for any $\{a_k\}_1^{n+1}$. Therefore

$$\|\textstyle\sum_{k=1}^{n} a_k\chi_k\|_E = \frac{1}{2}\|(\sum_{k=1}^{n} a_k\chi_k + a_{n+1}\chi_{n+1}) +$$

$$+ (\textstyle\sum_{k=1}^{n} a_k\chi_k - a_{n+1}\chi_{n+1})\| \le \|\sum_{k=1}^{n+1} a_k\chi_k\|_E. \quad \square$$

Remark 4.a.2. There is another way to prove Theorem 4.a.1. Equality (2.2) shows that

$$E_{\mathcal{F}_n}(\sum_{k=1}^{n+1} a_k\chi_k) = \sum_{k=1}^{n} a_k\chi_k. \qquad (4.a.2)$$

Consequently, Theorem 1.e.6 implies (4.a.1).

Theorem 4.a.3. *The H.s. is a monotone basis in any separable r.i. space.*

Proof. The H.s. monotonicity was proved in Theorem 4.a.1. The set of Haar polynomials is dence in any separable r.i. space. By Theorem 1.b.1. the H.s. is a monotone basis. □

Theorem 4.a.3. shows that

$$\|\sum_{k=i}^{j} a_k\chi_k\|_E \leq 2\|\sum_{k=1}^{\infty} a_k\chi_k\|_E \tag{4.a.3}$$

for every $1 \leq i \leq j < \infty$ and every r.i. space E.

Let as consider the majorant of the partial sums of Fourier-Haar series

$$S^{\#}x(t) = sup_{1\leq m<\infty} \mid S_m x(t) \mid .$$

Theorem 4.a.4. *Given* $p \in [1,\infty)$, $x \in L_p$, $u > 0$, *the following inequality*

$$mes\,\{t : t \in (0,1),\ S^{\#}x(t) > u\} \leq u^{-p} \int_{\{t:\, S^{\#}x(t)>u\}} |x(s)|^p\, ds \leq u^{-p}\|x\|_{L_p}^p$$

holds.

To prove Theorem 4.a.4. we introduce the following concept.

Definition 4.a.5. *A function* $\lambda(t)$ *mapping* $(0,1)$ *into* $\{1,2,...,\infty\}$ *is said to be stoping time if*

$$\{t : t \in (0,1),\ \lambda(t) = m\} \in \mathcal{F}_m \quad \forall m = 1,2,\ldots$$

Proof. Given $u > 0$, we put

$$\lambda(t) = \begin{cases} \min\{m : |S_m x(t)| > u\}, & \text{if } S^{\#}x(t) > u \\ \infty, & \text{if } S^{\#}x(t) \leq u. \end{cases}$$

For every $m = 2^n + k$, $(n,k) \in \Omega_1$

$$\{t : t \in (0,1), \lambda(t) = m\} \subset \{\emptyset, \Delta_n^k, \Delta_{n+1}^{2k-1}, \Delta_{n+1}^{2k}\}.$$

Therefore λ is the stopping time. As

$$\{t : t \in (0,1),\ \lambda(t) = m\} \subseteq \{t :\ t \in (0,1),\ |S_m x(t)| > u\} \quad \forall m = 1,2,...,$$

then

$$m\{t : t \in (0,1),\ \lambda(t) = m\} \leq u^{-p} \int_{\{t:\ \lambda(t)=m\}} |S_m x(s)|^p\, ds.$$

It follows from (2.4) and the Iensen inequality that

$$|S_m x|^p \leq E_{\mathcal{F}_m}|x|^p, \quad m = 1,2,...$$

Therefore

$$mes\{t \in (0,1): \quad S^{\#}x(t) > u\} = mes\{t : \lambda(t) < \infty\} =$$

$$= \sum_{m=1}^{\infty} mes\{t : \lambda(t) = m\} \le \sum_{m=1}^{\infty} u^{-p} \int_{\{t:\lambda(t)=m\}} | S_m x(s) |^p \, dt \le$$

$$\le \sum_{m=1}^{\infty} u^{-p} \int_{\{t:\lambda(t)=m\}} E_{\mathcal{F}m} | x |^p (s) \, ds =$$

$$= u^{-p} \sum_{m=1}^{\infty} \int_{\{t:\lambda(t)=m\}} | x(s) |^p \, ds =$$

$$= u^{-p} \int_{\{t:S^{\#}x(t)>u\}} | x(s) |^p \, ds \le u^{-p} \|x\|_{L_p}^p. \quad \square$$

Corollary 4.a.6. *The Fourier-Haar series of each function $x \in L_1$ converges to x a.e. If E is an r.i. space, $\beta_E < 1$, $x \in E$, then*

$$\|x\|_E \le \|S^{\#}x\|_E \le C\|x\|_E,$$

where C depends only on E.

Proof. Let n be an integer and $\varepsilon > 0$. By Theorem 4.a.4.,

$$mes\{t : sup_{m \ge n} | S_m x(t) - S_n x(t) | > \varepsilon\} \le \varepsilon^{-1} \|x - S_n x\|_{L_1}.$$

Theorem 4.a.3. states that the right part of the last inequality tends to 0. It means that $S_n x(t) \to x(t)$ a.e.

To prove the second part we must establish

Lemma 4.a.7. *Let E be an r.i. space, $\beta_E < 1$, $a > 0$, x and y be non-negative functions satisfying for any $u > 0$*

$$u \, mes\{t : t \in (0,1), \ x(t) > u\} \le a \int_{\{t:x(t)>u\}} y(s) \, ds \tag{4.a.4}$$

Then we have $\|x\|_E \le aC\|y\|_E$, where C depends only on E.

Proof. We prove that for each $t \in (0,1)$

$$x^*(t) \le a t^{-1} \int_0^t y^*(s) \, ds. \tag{4.a.5}$$

Recall that $\eta_x(u)$ is a distribution function of x. Fix $t \in (0,1)$ and suppose that $\eta_x(x^*(t)) = t$. Using assumption (4.a.4) we have

$$x^*(t) \le a t^{-1} \int_{\{s:\, x(s) > x^*(t)\}} y(s) \, ds.$$

Then (1.c.2) implies (4.a.5).

Now suppose that $\eta_x(x^*(t)) < t$. It means that $u_0 := x^*(t)$ is a discontinuity point of the function η_x. The function η_x is decreasing, therefore $\lim_{u \to u_0 - 0} \eta_x(u)$ exists. Denote it by t_1. Clearly,

$$t \le t_1, \quad x^*(t) = x^*(t_1) = u_0. \tag{4.a.6}$$

Let $0 < \varepsilon < u_0$. Using (1.c.2) and (4.a.4) we get

$$(x^*(t_1) - \varepsilon)\eta_x(u_0 - \varepsilon) \le a \int_0^{\eta_x(u_0 - \varepsilon)} y^*(s)\, ds.$$

Passing on to $\varepsilon \to 0$ we have

$$t_1 x^*(t_1) \le a \int_0^{t_1} y^*(s)\, ds. \tag{4.a.7}$$

Then (4.a.6) and (4.a.7) imply (4.a.5). So, inequality (4.a.5) is proved for any $t \in (0,1)$. By (1.d.6),

$$\|x\|_E \le a \int_0^1 \|\delta_{1/s}\|_E\, ds \|y\|_E. \quad \square$$

Now we are able to continue the proof of Collorary 4.a.6. The right part of inequality (4.a.3) follows from Theorem 4.a.4 and Lemma 4.a.7. The left part (4.a.3) follows from the first part proved above.

We don't use the condition $\beta_E < 1$ here. On the other hand, this condition is necessary for the validity of the inequality $\|S^{\#}x\|_E \le C\|x\|_E$, $x \in E$. This statement will be proved in Theorem 8.6.

4.b. Spaces $L_p(\omega)$

Recall that a function $\omega(t)$ on $[0,1]$ is named quasi-concave if $\omega(0) = 0$, $\omega(t)$ and $t/\omega(t)$ are increasing on $(0,1]$. Then $\omega(t)$ is a continuos function on $(0,1]$. We shall assume in addition that $\omega(t)$ is continuos in 0. Consider the function

$$\tilde{\omega}(t) = \inf f(t),$$

where the infinum runs over all concave functions f such that $f \ge \omega$. Then $\tilde{\omega}$ is concave and

$$\omega(t) \le \tilde{\omega}(t) \le 2\omega(t). \tag{4.b.1}$$

Each quasi-concave function ω generates a space $L_p(\omega)$ with the seminorm

$$\|x\|_{L_p(\omega)} = \sup_{0<\tau\le 1} \frac{\omega_p(\tau, x)}{\omega(\tau)},$$

where $\omega_p(\tau, x)$ is the continuity modulus of a function x in L_p. By (4.b.1) $L_p(\omega)$ and $L_p(\tilde{\omega})$ coinside up to equivalence. Therefore we shall assume that ω is concave. It is easy to check that for $\tau \geq 2^{-n-1}$

$$\omega_p(\tau, \chi_n^k) = \left(\int_0^1 \mid \chi_n^k(t + 2^{-n-1}) - \chi_n^k(t) \mid^p dt\right)^{\frac{1}{p}} =$$

$$= (2^{p-n-1} + 2^{-n})^{\frac{1}{p}} = 2^{-\frac{n}{p}}(2^{p-1} + 1)^{\frac{1}{p}}.$$

Recall that a function x is extended periodically on $[-1, 0)$ and $(1, 2]$ when we calculate $\omega_p(\tau, x)$.

If $0 < \tau < 2^{-n-1}$, $(n, k) \in \Omega_1$, then

$$\omega_p(\tau, \chi_n^k) = (\int_0^1 \mid \chi_n^k(t + \tau) - \chi_n^k(t) \mid^p dt)^{\frac{1}{p}} = (2^p\tau + 2\tau)^{\frac{1}{p}} = (2^p + 2)^{\frac{1}{p}}\tau^{\frac{1}{p}}.$$

Hence,

$$\|\chi_n^k\|_{L_p(\omega)} = (2^p + 2)^{\frac{1}{p}} \sup_{0 < \tau \leq 2^{-n-1}} \frac{\tau^{\frac{1}{p}}}{\omega(\tau)}.$$

Let x be a Haar polynomial of degree $\leq 2^n$, $0 < \tau \leq 2^{-n}$, $s > 0$. Clearly,

$$mes\{t : |x(t) - x(t + \tau)| > s\} = 2^n\tau \; mes\{t : |x(t) - x(t + 2^{-n})| > s\}.$$

Consequently,

$$(\int_0^1 \mid x(t) - x(t + \tau) \mid^p dt)^{\frac{1}{p}} = (2^n\tau)^{\frac{1}{p}}(\int_0^1 \mid x(t) - x(t + 2^{-n}) \mid^p dt)^{\frac{1}{p}}.$$

For each τ, x we get

$$\omega_p(\tau, x) = (2^n\tau)^{\frac{1}{p}}\omega_p(2^{-n}, x). \tag{4.b.2}$$

Theorem 4.b.1. *The H.s. is a basis in $L_p(\omega)$ iff there exists a constant $C > 0$ such that for every $0 < t_1 \leq t_2 \leq 1$*

$$\frac{t_1^{\frac{1}{p}}}{\omega(t_1)} \leq C\frac{t_2^{\frac{1}{p}}}{\omega(t_2)}. \tag{4.b.3}$$

Proof. Necessity. Consider the sequence of functions

$$x_n(t) = \begin{cases} \omega(t), & 0 \leq t \leq 2^{-n-1}; \\ \omega(2^{-n} - t), & 2^{-n-1} \leq t \leq 2^{-n}; \\ 0, & 2^{-n} \leq t \leq 1. \end{cases}$$

There exists a constant $C_1 > 0$ such that

$$\|S_{2^n}x_n\|_{L_p(\omega)} \leq C_1\|x_n\|_{L_p(\omega)} \quad \forall n = 1, 2, ... \tag{4.b.4}$$

As

$$\left(\int_0^1 | x_n(t+h) - x_n(t) |^p \, dt\right)^{\frac{1}{p}} \leq \begin{cases} 2^{-\frac{n-1}{p}}\omega(2^{-n-1}), & 2^{-n-1} \leq h \leq 1; \\ 2^{-\frac{n-1}{p}}\omega(h), & 0 \leq h \leq 2^{-n-1}; \end{cases}$$

then

$$\|x_n\|_{L_p(\omega)} \leq 2^{-\frac{n-1}{p}}. \tag{4.b.5}$$

Clearly,

$$S_{2^n}x_n(t) = \begin{cases} 2^{n+1}\int_0^{2^{-n-1}} \omega(s)\, ds, & 0 < t < 2^{-n}; \\ 0, & 2^{-n} < t < 1. \end{cases}$$

For $\tau \geq 2^{-n}$, we get

$$\omega_p(\tau, S_{2^n}x_n) = \omega_p(2^{-n}, S_{2^n}x_n) \quad \forall n = 1, 2, ...$$

By (4.b.2), we have

$$\|S_{2^n}x_n\|_{L_p(\omega)} = \sup_{0<\tau\leq 2^{-n}} \frac{(2^n\tau)^{\frac{1}{p}}\omega(2^{-n}, S_{2^n}x_n)}{\omega(\tau)} \geq$$

$$\geq 2^{n+1}\int_0^{2^{-n-1}} \omega(t)\, dt 2^{-\frac{n-1}{p}} \sup_{0<\tau\leq 2^{-n}} \frac{(2^n\tau)^{1/p}}{\omega(\tau)} \geq$$

$$\geq 2^{n+\frac{1}{p}+1}\int_{2^{-n-2}}^{2^{-n-1}} \omega(t)\, dt \sup_{0<\tau\leq 2^{-n}} \frac{\tau^{1/p}}{\omega(\tau)} \geq$$

$$\geq 2^{n+\frac{1}{p}+1} \cdot 2^{-n-2}\omega(2^{-n-2}) \sup_{0<\tau\leq 2^{-n}} \frac{\tau^{1/p}}{\omega(\tau)} \geq$$

$$\geq 2^{\frac{1}{p}-3}\omega(2^{-n}) \sup_{0<\tau\leq 2^{-n}} \frac{\tau^{1/p}}{\omega(\tau)}.$$

Using (4.b.5) and (4.b.4), we get

$$2^{\frac{1}{p}-3}\omega(2^{-n}) \sup_{0<\tau\leq 2^{-n}} \frac{\tau^{1/p}}{\omega(\tau)} \leq C_1 2^{-\frac{n-1}{p}}$$

and

$$\sup_{0<\tau\leq 2^{-n}} \frac{\tau^{1/p}}{\omega(\tau)} \leq 8C_1 \frac{2^{-\frac{n}{p}}}{\omega(2^{-n})}.$$

This inequality implies (4.b.3). We can choose $C = 16C_1$.

Sufficiency. Applying (2.4), we get

$$\int_0^1 \mid S_{2^n}x(t+2^{-n}) - S_{2^n}x(t) \mid^p \, dt \leq$$

$$\leq 2^{n(p-1)} \sum_{k=1}^{2^n} (\int_{\Delta_n}^k \mid x(t+2^{-n}) - x(t) \mid \, dt)^p \, dt \leq$$

$$\leq 2^{n(p-1)} \sum_{k=1}^{2^n} \int_{\Delta_n}^k \mid x(t+2^{-n}) - x(t) \mid^p \, dt 2^{-n(p-1)} =$$

$$= \int_0^1 \mid x(t+2^{-n}) - x(t) \mid^p \, dt \leq (\omega_p(2^{-n}, x))^p$$

Let $0 < \tau \leq 2^{-n}$. By (4.b.2) and (4.b.3), we have

$$\omega_p(\tau, S_{2^n}x) = (\tau 2^n)^{1/p} \omega_p(2^{-n}, S_{2^n}x) \leq$$

$$\leq (\tau 2^n)^{1/p} \omega_p(2^{-n}, x) \leq (\tau 2^n)^{1/p} \|x\|_{L_p(\omega)} \omega(2^{-n}) \leq$$

$$\leq C \|x\|_{L_p(\omega)} \omega(\tau).$$

Suppose that $m = 2^n + k$, $(n, k) \in \Omega_1$ and $0 < \tau \leq 2^{-n-1}$. Then

$$\omega_p(\tau, S_m x) \leq \omega_p(\tau, S_{2^n}x) + \omega_p(\tau, S_{2^{n+1}}x) \leq 2C \|x\|_{L_p(\omega)} \omega(\tau).$$

If $2^{-n-1} < \tau$, then $\frac{1}{m} < 2^{-n} < 2\tau$. By Theorems 3.1, 3.2,

$$\omega_p(\tau, S_m x) \leq \omega_p(\tau, x) + \omega_p(x - S_m x) \leq$$

$$\leq \omega_p(\tau, x) + 2\|x - S_m x\|_{L_p} \leq \omega_p(\tau, x) + (1 + 2^p)^{1/p} 2^{1+1/p} \omega_p(\tfrac{1}{m}, x) \leq$$

$$\leq (1 + (1 + 2^p)^{1/p} 2^{2+1/p}) \omega_p(\tau, x).$$

So,

$$\|S_m x\|_{L_p(\omega)} \leq \max(2C, 10) \|x\|_{L_p(\omega)}. \quad \square$$

Chapter 5

The Unconditionality of the Haar system

In order to characterize those r.i. spaces for which the Haar basis is unconditional, we study the operator

$$T_\varepsilon x = \sum_{m=1}^{\infty} \varepsilon_m c_m(x)\chi_m, \tag{5.1}$$

where $\varepsilon = \{\varepsilon_m\}_{m=1}^{\infty}$, $\varepsilon_m = \pm 1$.

Theorem 5.1. *Let $x \in L_1$. For every choice of signs ε the series (5.1) converges almost everywhere and for every $u > 0$*

$$mes\{t \in (0,1) : \sup_{N \geq 1} \mid \sum_{m=1}^{N} \varepsilon_m c_m(x)\chi_m(t) \mid \geq u\} \leq \frac{2}{u}\|x\|_{L_1}. \tag{5.2}$$

Proof. Observe first that it suffices to prove (5.2) only for polynomials with respect to the H.s. Indeed, if (5.2) is valid for polynomials, then, for every $u \geq 0$ and every M,

$$\begin{aligned} & mes\{t \in (0,1) : \sup_{N\geq 1} \mid \textstyle\sum_{m=M}^{M+N} \varepsilon_m c_m(x)\chi_m(t) \mid \geq u\} = \\ & = \lim_{N\to\infty} mes\{t \in (0,1) : \max_{1\leq N_1 \leq N} \mid \textstyle\sum_{m=M}^{M+N_1} \varepsilon_m c_m(x)\chi_m(t) \mid \geq u\} \leq \\ & \leq \tfrac{2}{u} \lim_{N\to\infty} \| \textstyle\sum_{m=M}^{M+N} c_m(x)\chi_m \|_{L_1} = \tfrac{2}{u} \| \textstyle\sum_{m=M}^{\infty} c_m(x)\chi_m \|_{L_1}. \end{aligned} \tag{5.3}$$

By virtue of Theorem 4.a.3 the last expression tends to zero if $M \to \infty$. This means that series (5.1) converges almost everywhere. It remains to remark that (5.3) coincides with (5.2) for $M = 1$. So, we shall prove (5.2) for polynomials with respect to the H.s. Fix $a, b \in R$ and $d \in [0,1]$ and let $P_M(a,b,d)$ be the set of polynomials of order M, $M = 2, 3, \ldots$, with respect to the H.s.

$$x(t) = a + \sum_{m=2}^{M} c_m\chi_m(t),$$

for which there exists a choice of signs $\{\varepsilon_m\}_{m=2}^M$, $\varepsilon_m = \pm 1$ such that

$$mes\{t \in (0,1) : \max_{1\le k\le M} | a + \sum_{m=2}^{k} \varepsilon_m c_m \chi m(t) - b | \ge 1\} \ge d.$$

Put $h_M(a,b,d) := \min\{\|x\|_{L_1} : \quad x \in P_M(a,b,d)\}$. It is easily verified that inequality (5.2) will be proved if we establish that, for every $M = 2, 3, ...,$

$$h_M(0,0,d) \ge \frac{d}{2}, \quad d \in [0,1]. \tag{5.4}$$

In order to prove (5.4) we shall construct the function $g(a,b,d)$, $a, b \in R$, $d \in [0,1]$ such that $g(0,0,d) \ge \frac{d}{2}$, $d \in [0,1]$ and, for every $M = 2, 3, ...$

$$h_M(a,b,d) \ge g(a,b,d), \quad a, b, \in R, \quad d \in [0,1]. \tag{5.5}$$

Let us find under what conditions on g it is possible to check (5.5) by induction on M. The function h_2 is calculated directly:
1) $h_2(a,b,0) = | a |$, $a, b \in R$;
2) put $A := \{(a,b) : -1 \le b \le 1; \quad 2a - 1 \le b \le 2a + 1\}$,
if $d \in (0, 1/2)$, then

$$h_2(a,b,d) = \begin{cases} | a |, & (a,b) \notin A; \\ 1 - | a - b |, & (a,b) \in A; \end{cases}$$

if $d \in [\frac{1}{2}, 1]$, then

$$h_2(a,b,d) = \begin{cases} | a |, & (a,b) \notin A; \\ 1 + | a - b |, & (a,b) \in A. \end{cases}$$

So, the first condition on g is the following one

$$g(a,b,d) \le h_2(a,b,d), \quad a, b \in R, \quad d \in [0,1]. \tag{5.6}$$

Suppose now that inequality (5.5) is valid for some M. Consider the polynomial

$$x = a + \sum_{m=2}^{M+1} c_m \chi_m,$$

belonging to $P_{M+1}(a,b,d)$, $a, b \in R$, $d \in [0,1]$. Hence, there exists the choice of signs $\varepsilon = \{\varepsilon_m\}_{m=2}^{M+1}$ such that

$$mes\{t \in (0,1) : \max_{2\le N\le M+1} | a + \sum_{m=2}^{N} \varepsilon_m c_m \chi_m(t) - b | \ge 1\} \ge d.$$

For $j = 0, 1$ put

$d_j := mes\{t \in (\frac{j}{2}, \frac{j+1}{2}) : \max_{2\le N\le M+1} | a + \sum_{m=2}^{N} \varepsilon_m c_m \chi_m(t) - b | \ge 1\}$,
$x_j(t) = x(\frac{j}{2} + \frac{t}{2})$, $t \in (0,1)$.

It is clear that

$$x_j \in P_M(a+(-1)^j c_2, b+(-1)^j(1-\varepsilon_2)c_2, 2d_j).$$

Thus, by applying the induction hypothesis, we get that

$$\begin{aligned} \|x\|_{L_1} &= 2^{-1}(\|x_0\|_{L_1} + \|x_1\|_{L_1}) \geq \\ &\geq 2^{-1}(g(a+c_2, b+(1-\varepsilon_2)c_2, 2d_0) + g(a-c_2, b-(1-\varepsilon_2)c_2, 2d_1). \end{aligned}$$

This means that inequality (5.5) can be checked by induction on M if the function g satisfies the following conditions:
1) for every $a, b, c \in R$ and every $d_0, d_1 \in [0, \frac{1}{2}]$.

$$2^{-1}(g(a+c, b, 2d_0) + g(a-c, b, 2d_1)) \geq g(a, b, d_0+d_1);$$

$$2^{-1}(g(a+c, b+2c, 2d_0) + g(a-c, b-2c, 2d_1)) \geq g(a, b, d_0+d_1);$$

2) if $0 \leq d_0 \leq d_1 \leq 1$ then

$$g(a, b, d_0) \leq g(a, b, d_1), \quad a, b \in R.$$

The following function

$$g(a,b,d) := \begin{cases} \max(|\, a \,|, \frac{d}{2} + ab - \frac{b^2}{2}), & \max(|\, a \,|, |\, b \,|) < 1; \\ |\, a \,|, & \max(|\, a \,|, |\, b \,|) \geq 1. \end{cases}$$

satisfies all the above mentioned conditions. Taking into account the previous arguments, we achieve the proof of Theorem 5.1. □

Remark 5.2. We can guess the kind of the formula for g by the following reasoning. Fix $b_0 \in [-1, 1]$ and put $f(a) := h_2(a, b_0, 2^{-1})$, $a \in R$. By (5.6) we get that $g(a, b_0, 2^{-1}) \leq f(a)$. Besides, condition 1) means, in particular, the convexity of g along the line $b = b_0$, $d = 2^{-1}$. Let us "correct" the function f on the interval $[2^{-1}(b_0-1), 2^{-1}(b_0+1)]$ by continuing it linearly from the ends of the interval. We get the new function

$$f_1(a) = \max(|\, a \,|, \frac{1}{2} + ab_0 - \frac{b_0^2}{2}), \quad a \in R.$$

It suggestes the formula for g.

Remark 5.3. *The constant 2 in (5.2) is the best possible one.*

Consider the function

$$x = 2^{-1}\chi_1 + 2^{-1}\chi_2 + \chi_3$$

and put $\varepsilon_1 = \varepsilon_3 = 1, \quad \varepsilon_2 = -1$. In this case we have $\|x\|_{L_1} = 2^{-1}$ and

$$mes\{t \in (0,1) : | T_\varepsilon x(t) | \geq 1\} = 1.$$

More general results are proved in [47]. Let Φ be an increasing convex function on $[0,\infty)$ such that Φ is twice differentiable on $(0,\infty)$; Φ' is strictly concave, and $\Phi(0) = \Phi'(0+) = 0$. Then, for every $u > 0$

$$mes\{t \in (0,1) : \sup_{N\geq 1} | \sum_{m=1}^{N} \varepsilon_m c_m(x)\chi_m(t) | \geq u\} \leq$$
$$\leq 2\left(\int_0^\infty \Phi(s)e^{-s}ds)^{-1} \int_0^1 \Phi(u^{-1} \mid x(t) \mid)\right) dt,$$

where $\varepsilon_m = \pm 1, \quad m = 1,2,\ldots$ The constant on the right is the best possible one. In particular, for $x \in L_p, \quad 1 \leq p \leq 2$, we have

$$\sup_{u>0} u^p mes\{t \in (0,1) : \sup_{N\geq 1} | \sum_{m=1}^{N} \varepsilon_m c_m \chi_m(t) | \geq u\} \leq \frac{2}{\Gamma(p+1)} \|x\|_{L_p}^p,$$

where Γ is the Gamma-function.

Inequality (5.2) and its generalizations are valid for arbitrary sequences of the following form $\{\varepsilon_m\}_{m=1}^\infty, \quad | \varepsilon_m | \leq 1$ [47].

Remark 5.4. There exist some choices of signs $\varepsilon = \{\varepsilon_m\}_{m=1}^\infty$ for which the operator T_ε is not bounded in L_1. For example, if

$$x_N := \chi_0^1 + \sum_{n=1}^{N} 2^{n-1}\chi_n^1, \quad N = 1,2,\ldots,$$

then $\|x_N\|_{L_1} = 1$, but

$$\lim_{N\to\infty} \|\chi_0^1 + \sum_{n=1}^{N} (-1)^n 2^{n-1} \chi_n^1\|_{L_1} = \infty.$$

Thus, the Haar basis in $L_1(0,1)$ is not unconditional. Incidentally, $L_1(0,1)$ fails to have an unconditional basis [225]. V.M. Kadec and M.M. Popov proved in [130] that the H.s. is not unconditional in any hyperoctant of $L_1(0,1)$. By the hyperoctant corresponding to a choice of signs $\varepsilon_k = \pm 1, \quad k = 1,2,\ldots$, we mean a set of functions $x = \sum_{k=1}^\infty \varepsilon_k a_k \chi_k$ with positive a_k.

Theorem 5.5. *Let* $1 < p < \infty, \quad p^* = \max(p, \frac{p}{p-1})$. *For every function* $x \in L_p$ *and every* $\varepsilon = \{\varepsilon_n\}_{n=1}^\infty, \quad \varepsilon_n = \pm 1$,

$$\frac{1}{p^*-1}\|x\|_{L_p} \leq \|T_\varepsilon x\|_{L_p} \leq (p^*-1)\|x\|_{L_p}. \tag{5.7}$$

Proof. Define the functions z_n, $n \geq 1$, on the interval [0,1] with values in R^2

$$z_n(t) := (T_\varepsilon S_n x(t) + S_n x(t), T_\varepsilon S_n x(t) - S_n x(t)).$$

Let

$$f(s,t) := \mid (s+t)/2 \mid^2 - (p^* - 1)^p \mid (s-t)/2 \mid^p, \quad s,t \in R.$$

It is clear that

$$\int_0^1 f(z_n(w))dw = \|T_\varepsilon S_n x\|_{L_p}^p - (p^* - 1)^p \|S_n x\|_{L_p}^p.$$

By theorem 4.a.3 the right inequality in (5.7) is equivalent to the following

$$\int_0^1 f(z_n(w))dw \leq 0, \quad n = 1, 2, \dots . \tag{5.8}$$

The idea of the proof of (5.8) consists of the construction and employment of the function g with the folowing properties :
1) g is concave both along the first variable and along the second one;
2) $f \leq g$;
3) if $st = 0$, then $g(s,t) \leq 0$.

Let us show first how the function g is used to prove (5.8). Since, $(T_\varepsilon S_1 x + S_1 x)(T_\varepsilon S_1 x - S_1 x) = 0$, then by 3) we have $\int_0^1 g(z_1(w))dw \leq 0$. Notice now that for every $n = 1, 2, \dots$

$$\int_0^1 g(z_n(w))dw \leq \int_0^1 g(z_{n-1}(w))dw. \tag{5.9}$$

Indeed, if $uv = 0$ and the partial derivatives $g_s := \frac{\partial g}{\partial s}$ and $g_t := \frac{\partial g}{\partial t}$ exist, then by 1) we obtain

$$g(s+u, t+v) \leq g(s,t) + g_s(s,t)u + g_t(s,t)v \tag{5.10}$$

Therefore, using the trivial equality

$$(T_\varepsilon S_n x + S_n x - T_\varepsilon S_{n-1} x - S_{n-1} x)(T_\varepsilon S_n x - S_n x - T_\varepsilon S_{n-1} x + S_{n-1} x) =$$

$$= (\varepsilon_n c_n(x)\chi_n + c_n(x)\chi_n)(\varepsilon_n c_n(x)\chi_n - c_n(x)\chi_n) = 0,$$

we get that

$$g(z_n) \leq g(z_{n-1}) + g_s(z_{n-1})(\varepsilon_n c_n(x)\chi_n + c_n(x)\chi_n) +$$

$$+ g_t(z_{n-1})(\varepsilon_n c_n(x)\chi_n - c_n(x)\chi_n). \tag{5.11}$$

Remark now that

$$\int_0^1 g_s(z_{n-1}(w))(\varepsilon_n c_n(x) + c_n(x))\chi_n(w)dw =$$

$$= \int_0^1 g_t(z_{n-1}(w))(\varepsilon_n c_n(x) - c_n(x))\chi_n(w)dw = 0$$

since the functions $g_s(z_{n-1}(w))$ and $g_t(z_{n-1}(w))$ are constant on the $supp\,\chi_n$. By integrating inequality (5.11) over the interval [0,1] and taking into account the last equations, we obtain (5.9). By property 2), (5.9) implies (5.8).

So, it only remains to construct the function g. If $p = 2$, then $f(s,t) = st$ and in this case we can put $g := f$. Let $p > 2$. Consider the following sets

$$A_1 := \{(s,t) \in R^2 : s > 0, \quad (1 - \tfrac{2}{p^*})s < t < s\},$$
$$A_2 := \{(s,t) \in R^2 : s > 0, \quad -s < t < (1 - \tfrac{2}{p^*})s)\}.$$

Let $\alpha_p = p(p^*/(p^* - 1))^{1-p}$. Define the function g by the formula

$$g(s,t) = \begin{cases} \alpha_p s^p \left(1 - \frac{p^*(s-t)}{2s}\right), & (s,t) \in A_1; \\ f(s,t), & (s,t) \in A_2. \end{cases}$$

For the rest (s,t) the function g is determined according to the equations $g(s,t) = g(t,s) = g(-s,-t), \quad (s,t) \in R^2$. By direct calculations, it is checked that the partial derivatives g_s and g_t exist and are continuous on R^2. Moreover, outside of the lines $s+t=0, \quad s=t, \quad t=(1-\frac{2}{p^*})s, \quad s=(1-\frac{2}{p^*})t$ the second partial derivatives satisfy the inequalities : $g_{ss} \le 0, \quad g_{tt} \le 0, \quad g_{st} \ge 0$. Fix s,t,u and v such that $uv \le 0, \quad u+v \ne 0$. Define the function $h(r) := g(s+ur, t+vr), \quad r \in R$. Except the finite number of values of r, the function g is twice continuously differentiable in the neighbourhood of $(s+ur, t+vr)$ and

$$h''(r) = u^2 g_{ss}(s+ur,t+vr) + 2uvg_{st}(s+ur,t+vr) + v^2 g_{tt}(s+ur,t+vr) \le 0.$$

Since h' is continuous, the last inequality means that h is concave. Therefore, (5.10) is true for every u and v, $uv \le 0$.

Let us prove now that $f \le g$. We need to prove this only for A_1. Fix $s > 0$ and put $a = (1 - \frac{2}{p^*})s, \quad b = s$. Define the function $e(t) = g(s,t) - f(s,t), \quad t \in [a,b]$. It is easy to show that e' is continuous on $[a,b]$ and $e''(t) = -f_{tt}(s,t), \quad t \in [a,b]$. In addition,

$$e(a) = 0, \quad e(b) = \alpha_p b^p - b^p > 0, \tag{5.12}$$

and there exists an $a_1 \in (a,b)$ such that $e''(t) \ge 0$ on $(0,a_1]$ and $e''(t) \le 0$ on (a_1,b). Since $e'(a) = 0$, then $e'(t) \ge 0$ on $[a,a_1]$. By (5.12), we get that $e(t) \ge 0$ on $[a,a_1]$. On $[a_1,b]$ the function e is concave and has positive values in the ends of the interval. That is why $e(t) \ge 0$ on $[a_1,b]$. Thus, $f \le g$ on A_1 and therefore on R^2.

So, the right inequality of (5.7) is proved for $p > 2$. Since $T_\varepsilon(T_\varepsilon x) = x$, then

$$\|x\|_{L_p} = \|T_\varepsilon(T_\varepsilon x)\|_{L_p} \le (p-1)\|T_\varepsilon x\|_{L_p}.$$

For $p \in (1,2)$ the proof of the theorem can be obtained either by duality or by similar arguments if we interchange the sets A_1 and A_2 in the definition of g.

Remark 5.6. The constant $(p^* - 1)$ in (5.7) is the best possible one. The proof of this will be given in Remark 8.5.

Remark 5.7. Not being interested in the best possible constant we can give another proof of Theorem 5.5. Accoding to Theorem 5.1 the operator T_ε is well defined on L_1 and has the weak type (1,1). Besides, it is obvious that T_ε is an isometry in L_2. Therefore, by the Marcinkiewicz interpolation theorem for $p \in (1,2)$, there exists a constant $C > 0$ such that

$$\|T_\varepsilon x\|_{L_p} \le \frac{Cp^2}{p-1}\|x\|_{L_p}, \quad x \in L_p.$$

Corollary 5.8. *The H.s. is an unconditional basis of L_p, $1 < p < \infty$, and*

$$UC_{L_p}(\{\chi_m\}_{m=1}^\infty) = p^* - 1.$$

This follows from Theorem 5.5 and remark 5.6.

Corollary 5.9. *Let J be a subset of integers. Define the projection*

$$P_J\chi_m = \begin{cases} \chi_m, & m \in J; \\ 0, & m \notin J. \end{cases}$$

Then,

$$\|P_J\|_{L_p} \le \frac{p^*}{2}. \tag{5.13}$$

Inequality (5.13) follows from (5.7) since $P_J = \frac{1}{2}(T_\varepsilon + I)$ with $\varepsilon = \{\varepsilon_m\}_{m=1}^\infty$ defined by

$$\varepsilon_m = \begin{cases} +1, & m \in J; \\ -1, & m \notin J; \end{cases}$$

where I is an identity.

Chapter 6

The Paley Function

For the further investigation of the unconditionality of the H.s. we study the properties of the operator which transforms the function $x \in L_1$ into the function

$$Px = \sum_{m=1}^{\infty} (c_m(x)\chi_m)^2)^{1/2}, \tag{6.1}$$

that is called the Paley function.

Theorem 6.1. *For every $x \in L_1$ the Paley function is finite almost everywhere on [0,1] and*

$$mes\,\{t \in (0,1) : Px(t) > u\} \leq \frac{C}{u}\|x\|_{L_1}, \quad u > 0, \tag{6.2}$$

with $C = e^{-1/2} + \int_0^1 e^{-t^2/2}\,dt$.

Proof. Since $PS_N x \uparrow Px$, when $N \to \infty$, and $\|S_N x\|_{L_1}$, $N = 1, 2, ...$, then it suffices to prove inequality (6.2) only for polynomials with respect to the H.s. Fix $a \geq 0$ and $b \geq 0$. Let $f_M(a,b)$, $M = 2, 3, ...$, be the infimum of L_1 - norms of the polynomials

$$x = a + \sum_{m=2}^{M} c_m \chi_m,$$

with the Paley function $Px \equiv (a^2 + b^2)^{1/2}$. It is obvious that $f_2(a,b) = \max(a,b)$. As in Theorem 5.1, we are going to prove by induction on M the following inequalities

$$f_M(a,b) \geq g(a,b), \quad a \geq 0, \quad b \geq 0, \quad M = 2, 3, ... \tag{6.3}$$

with some special function g. What properties of g are necessary for the realization of this idea?

The first property is

$$g(a,b) \le f_2(a,b), \quad a \ge 0, \quad b \le 0.$$

Suppose now that inequality (6.3) is valid for some M. Consider the polynomial

$$x = a + \sum_{m=2}^{M+1} c_m \chi_m,$$

with $Px \equiv (a^2+b^2)^{1/2}$. For $j = 0,1$ define

$$x_j(t) = x(\frac{j+t}{2}), \quad t \in (0,1).$$

It is clear that x_0 and x_1 are polynomials of the order not greater than M and

$$Px_j \equiv ((a+(-1)^j c_2)^2 + (b^2 - c_2^2))^{1/2}, \quad j = 0,1.$$

It is obvious that

$$\|x\|_{L_1} = 2^{-1}(\|x_0\|_{L_1} + \|x\|_{L_1}).$$

Thus, by applying the induction hypothesis, we get that

$$\|x\|_{L_1} \ge 2^{-1}(g(|\, a+c_2 \,|, (b^2-c_2^2)^{1/2}) + g(|\, a-c_2 \,|, (b^2-c_2^2)^{1/2}).$$

This means that inequality (6.3) can be checked by induction on M if the function g satisfies the following condition: for every $a \ge 0, \quad b \ge 0$ and $c, \quad |\, c \,| \le b,$

$$2^{-1}(g(|\, a+c \,|, (b^2-c^2)^{1/2}) + g(|\, a-c \,|, (b^2-c^2)^{1/2})) \ge g(a,b). \qquad (6.4)$$

The function

$$g(a,b) = \begin{cases} bh(\frac{a}{b})/h(1), & 0 \le a < b; \\ a, & 0 \le b \le a; \end{cases}$$

with $h(t) := e^{-t^2/2} + t\int_0^t e^{-s^2/2}\,ds$ satisfies all the above mentioned conditions. The verification of the last statement is rather tedious, that is why we give only the scheme of reasoning.

It is clear that it sufficies to prove (6.3) only for $b = 1$ and $a \in [0,1)$. So, instead of (6.4) we shall prove that for every $a \in [0,1)$ and $c \in [0,1)$

$$g(a,1) \le 2^{-1}(g(a+c,(1-c^2)^{1/2}) + g(|\, a-c \,|, (1-c^2)^{1/2})). \qquad (6.5)$$

First of all (6.5) is checked for $c \le a \le (1-c^2)^{1/2} - c$. If $c = 0$ and $a \in [0,1]$, inequality (6.5) is obviously valid. Besides, the partial derivative along c of the right part is positive. On the set $\{(a,c) : 0 \le a \le \min(c,(1-c^2)^{1/2} - c)\}$ the partial derivative along c of the right part of (6.5) is still positive. So, inequality (6.5) is valid on this set. Now we can consider the set $\{(a,c) : \max(c,(1-c^2)^{1/2}-c) < a \leq 1\}$. On it inequality (6.5) takes the form

$$\frac{h(a)}{h(1)} \leq \frac{1}{2}(a+c+\frac{(1-c^2)^{1/2}}{h(1)}h\left(\frac{a-c}{(1-c^2)^{1/2}}\right)). \tag{6.6}$$

The second partial derivative along c of the right part is non-positive on the set under consideration. That is why it sufficies to check (6.6) only for $a=c$. If $a=c$, then (6.6) takes the form

$$2h(c) \leq 2ch(1)+(1-c^2)^{1/2}, \quad c \in [5^{-1/2},1].$$

The last inequality is valid because it is true on the ends of the interval, and the second derivative of the difference of the right and the left parts of it is non-positive.

Consider the set $\{(a,c) : 0 \leq a \leq c-(1-c^2)^{1/2}, \quad c \leq 1\}$ on which inequality (6.5) is written in the form

$$\frac{h(a)}{h(1)} \leq c. \tag{6.7}$$

Since h is a monotone non-decreasing function, then (6.7) will be proved, if we check that

$$h(c-(1-c^2)^{1/2}) \leq ch(1), \quad c \in [2^{-1/2},1]. \tag{6.8}$$

Inequality (6.8) is true on the ends of the interval, and the second derivative of the left part is non-negative. Thus, (6.8) is valid for every $c \in [2^{-1/2},1]$.

It remains to consider the set $\{(a,c) :\mid c-(1-c^2)^{1/2} \mid\leq a \leq c\}$. Inequality (6.5) is already proved on the boundary of this set. So, it suffices to remark that the second partial derivative along c of the right part of (6.5) is non-positive. This reasoning, at last, completes the presentation of the scheme of the proof of (6.4) and together with it of the whole theorem 6.1. □

Remark 6.2. The approximate value of the constant C is equal to 1.4622. We do not know if it is the best possible or not. In paper [22] it is written that the best possible constant in inequality (6.2) is not less than 1.44. About the constant C see also Remark 8.10.

Theorem 6.3. *Let E be an r.i. space with $\alpha_E > 0$. For every $x \in E$,*

$$c_2(E)\|Px\|_E \leq \|S^{\#}x\|_E \leq c_1(E)\|Px\|_E. \tag{6.9}$$

To prove this theorem we need the following

Lemma 6.4. *Let E be an r.i. space. If two non-negative functions x and y from E satisfy for every $u > 0$ the inequality*

$$mes\,\{t \in (0,1) : x(t) > au, y(t) \le bu\} \le c\,mes\,\{t \in (0,1) : x(t) > u\} \quad (6.10)$$

with some positive numbers a, b, c, and there exists a $\gamma > 1$ such that $a\|\sigma_{\gamma c}\|_E < 1$, then

$$\|x\|_E \le \frac{a\|\sigma_{\gamma/(\gamma-1)}\|_E}{b(1 - a\|\sigma_{\gamma c}\|_E)}\|y\|_E. \quad (6.11)$$

Proof. Inequality (6.10) implies that

$$\begin{aligned} &mes\,\{t \in (0,1) : x(t) > u\} \le \\ &\le mes\,\{t \in (0,1) : x(t) > u, y(t) \le \tfrac{b}{a}u\} + mes\,\{t \in (0,1) : y(t) > \tfrac{b}{a}u\} \le \\ &\le c\,mes\,\{t \in (0,1) : x(t) > \tfrac{u}{a}\} + mes\,\{t \in (0,1) : y(t) > \tfrac{b}{a}u\}. \end{aligned}$$

This means that $\eta_x(u) \le c\eta_x(u/a) + \eta_y(bu/a)$ for every $u > 0$. That is why for $t < \min(c\gamma, \gamma/(\gamma-1))$ we get

$$\begin{aligned} x^*(t) &= \inf\{u > 0 : \eta_x(u) < t\} \le \inf\{u > 0 : c\eta_x(\tfrac{u}{a}) + \eta_y(\tfrac{bu}{a}) < t\} \le \\ &\le \inf\{u > 0 : c\eta_x(\tfrac{u}{a}) < \tfrac{t}{\gamma}, \eta_y(\tfrac{bu}{a}) < \tfrac{\gamma-1}{\gamma}t\} \le \\ &\le \max(\inf\{u : c\eta_x(\tfrac{u}{a}) < \tfrac{t}{\gamma}\}, \inf\{u : \eta_y(\tfrac{bu}{a}) < \tfrac{\gamma-1}{\gamma}t\}) = \\ &= \max(a\inf\{u : c\eta_x(u) < \tfrac{t}{\gamma}\}, \tfrac{a}{b}\inf\{u : \eta_y(u) < \tfrac{\gamma-1}{\gamma}t\}) = \\ &= \max(ax^*(\tfrac{t}{c\gamma}), \tfrac{a}{b}y^*(\tfrac{\gamma-1}{\gamma}t)) \le ax^*(\tfrac{t}{c\gamma}) + \tfrac{a}{b}y^*(\tfrac{\gamma-1}{\gamma}t). \end{aligned}$$

Hence, we have that

$$\|x\|_E \le a\|\sigma_{\gamma c}x\|_E + \frac{a}{b}\|\sigma_{\gamma/(\gamma-1)}y\|_E \le a\|\sigma_{\gamma c}\|_E\|x\|_E + \frac{a}{b}\|\sigma_{\gamma/(\gamma-1)}\|_E\|y\|_E.$$

The last inequality is equivalent to (6.11). □

Proof of Theorem 6.3. It suffices to check (6.9) only for polynomials with respect to the H.s. since

$$\begin{aligned} &S^{\#}x(t) = \lim_{N\to\infty} S^{\#}S_N x(t), \quad t \in (0,1); \\ &Px(t) = \lim_{N\to\infty} PS_N x(t), \quad t \in (0,1). \end{aligned}$$

So, later on we assume that

$$x = S_N x = \sum_{m=1}^{N} c_m \chi_m$$

for some $N = 1, 2, \ldots$ We prove first the right part of (6.9). Introduce the following notations:

$$d_k = c_k(x)\chi_k, \quad d_0 = e_0 = 0, \quad e_k = \max_{1\le i\le k} |\, d_i \,|,$$

$$A_k = \{t \in (0,1) : | d_k(t) | \leq 2e_{k-1}(t)\},$$

$$f_k = d_k \kappa_{A_k}, \quad g_k = d_k - f_k, \quad y_k = \sum_{i=1}^{k} f_i,$$

$$z_k = \sum_{i=1}^{k} g_i, \quad k = 1, 2, ..., N.$$

It is clear that $A_k \in \mathcal{F}_{k-1}$, $k = 1, 2, ..., N$, $(\mathcal{F}_0 := \mathcal{F}_1)$. That is why f_k is equal either to d_k or to 0. Remark the obvious inequalities

$$S^{\#}x \leq S^{\#}y_N + S^{\#}z_N \leq S^{\#}y_N + \sum_{k=1}^{N} | g_k |, \tag{6.12}$$

$$Py_N \leq Px_N + Pz_N \leq Px_N + \sum_{k=1}^{N} | g_k | . \tag{6.13}$$

Let us prove that for $a > 1$, $0 < b < a - 1$ and for every $u > 0$ the following inequality is true

$$mes\,\{t \in (0,1) : S^{\#}y_N(t) > au \max(Py_N(t), 2e_N(t)) \leq bu\} \leq$$
$$\leq 2b^2(a - b - 1)^{-2} mes\,\{t \in (0,1) : S^{\#}y_N(t) > u\}. \tag{6.14}$$

Fix $u > 0$ and put

$$\eta(t) = \begin{cases} \min\{k : | y_k(t) | > u\}, & S^{\#}y_N(t) > u; \\ \infty, & S^{\#}y_N(t) \leq u; \end{cases}$$

$$\lambda(t) = \begin{cases} \min\{k : | y_k(t) | > au\}, & S^{\#}y_N(t) > au; \\ \infty, & S^{\#}y_N(t) \leq au; \end{cases}$$

$$\mu(t) = \begin{cases} \min\{k : \max(Py_k(t), 2e_k(t)) > bu\}, & \max(Py_N(t), 2e_N(t)) > bu; \\ \infty, & \max(Py_N(t), 2e_N(t)) \leq bu; \end{cases}$$

$$B_k = \{t \in (0,1) : \eta(t) < k \leq \min(\lambda(t), \mu(t))\},$$

$$\tilde{y}_N = \sum_{k=1}^{N} f_k \kappa_{B_k}.$$

It follows immediately from the definitions that

$$B_k = \{t \in (0,1) : S^{\#}y_{k-1}(t) > u, \quad S^{\#}y_{k-1}(t) \leq au,$$
$$\max(Py_{k-1}(t), 2e_{k-1}(t)) \leq bu\} \in \mathcal{F}_{k-1}, \quad k = 1, 2, ..., N.$$

Therefore, $\tilde{y}_N$ is the polynomial of the order not greater than N with respect to the H.s. If $\eta(t) = \infty$, then $P\tilde{y}_N(t) = 0$. Besides, for every $t \in (0,1)$, $P^2\tilde{y}_N \leq 2b^2u^2$. Indeed, if $\mu(t) < \infty$, then

$$P^2\tilde{y}_N(t) \leq \sum_{k=1}^{\mu(t)} f_k^2(t) \leq \sum_{k=1}^{\mu(t)-1} f_k^2(t) + d^2_{\mu(t)}(t)\kappa_{A_{\mu(t)}}(t) \leq$$

$$\leq b^2u^2 + 2e^2_{\mu(t)-1}(t) \leq 2b^2u^2.$$

If $\mu(t) = \infty$, then

$$P^2\tilde{y}_N(t) \leq P^2 y_N(t) \leq b^2u^2.$$

By applying the above inequalities, we have that

$$\begin{aligned} \|\tilde{y}_N\|_2^2 &= \|P\tilde{y}_N\|_2^2 \leq 2b^2u^2 mes\,\{t \in (0,1) : \eta(t) < \infty\} = \\ &= 2b^2u^2 mes\,\{t \in (0,1) : S^{\#}y_N(t) > u\}. \end{aligned} \tag{6.15}$$

In addition, if $S^{\#}y_N(t) > au$, $\max(Py_N(t), 2e_N(t)) \leq bu$, then $\lambda(t) \leq N$, $\mu(t) = \infty$, and

$$\begin{aligned} &S^{\#}\tilde{y}_N(t) \geq \Big| \textstyle\sum_{k=\eta(t)+1}^{\lambda(t)} f_k(t) \Big| = \Big| y_{\lambda(t)} - y_{\eta(t)-1}(t) - f_{\eta(t)}(t) \Big| \geq \\ &\geq | y_{\lambda(t)}(t) | - | y_{\eta(t)-1}(t) | - | f_{\eta(t)}(t) | \geq (a-1-b)u. \end{aligned}$$

Thus,

$$\{t \in (0,1) : S^{\#}y_N(t) > au, \max(Py_N(t), 2e_N(t)) \leq bu\}$$

$$\subset \{t \in (0,1) : S^{\#}\tilde{y}_N(t) > (a-1-b)u\}.$$

To prove inequality (6.14) it remains to use Theorem 4.a.4

$$mes\,\{t \in (0,1) : S^{\#}y_N(t) > au;\ \max(Py_N(t), 2e_N(t)) \leq bu\} \leq$$

$$\leq mes\,\{t \in (0,1) : S^{\#}\tilde{y}_N(t) > (a-1-b)u\} \leq (a-1-b)^{-2}u^{-2}\|\tilde{y}_N\|_2^2 \leq$$

$$\leq 2b^2(a-b-1)^{-2} mes\,\{t \in (0,1) : S^{\#}y_N(t) > u\}.$$

Here, the last inequality is based on (6.15). The condition $\alpha_E > 0$ is equivalent to the following: $\lim_{s\to 0} \|\sigma_s\|_E = 0$. Therefore, we may use lemma 6.4, since inequality (6.14) is valid for every $b \in (0, a-1)$. So,

$$\|S^{\#}y_N\|_E \leq C_1(E)(\|Py_N\|_E + \|e_N\|_E).$$

By combining this inequality and (6.12), and (6.13), we obtain

$$\|S^{\#}x\|_E \leq C_1(E)\left(\|Px\|_E + \|e_N\|_E + \Big\|\sum_{k=1}^{N} | g_k | \Big\|_E\right). \tag{6.16}$$

Let us prove that

$$\sum_{k=1}^{N} | g_k | \le 2e_N. \tag{6.17}$$

Indeed, if $t \notin A_k$, then $| d_k(t) | > 2e_{k-1}(t)$, and therefore,

$$| d_k(t) | + 2e_{k-1}(t) \le 2 | d_k(t) | \le 2e_k(t),$$

that is

$$| g_k | \le 2(e_k - e_{k-1}), \quad k = 1, 2, ..., N.$$

Inequalities (6.16), (6.17) together with the obvious $e_N \le Px$ complete the proof of the right part of (6.9).

To prove the left part of (6.9), we show first that for $a > 1$, $0 < b^2 < a^2 - 1$ and every $u > 0$

$$mes\,\{t \in (0,1) : Py_N(t) > au,\ \max(S^{\#}y_N, 2e_N) \le bu\} \le$$

$$\le 16b^2(a^2 - b^2 - 1)^{-1} mes\,\{t \in (0,1) : Py_N(t) > u\}. \tag{6.18}$$

For fixed $u > 0$ we put

$$\eta'(t) = \begin{cases} \min\{k : Py_k(t) > u\}, & Py_N(t) > u; \\ \infty, & Py_N(t) \le u; \end{cases}$$

$$\lambda'(t) = \begin{cases} \min\{k : Py_k(t) > au\}, & Py_N(t) > au; \\ \infty, & Py_N(t) \le au; \end{cases}$$

$$\mu'(t) = \begin{cases} \min\{k : \max(S^{\#}y_k(t), 2e_k(t)) > bu\}, & \\ \qquad\qquad \text{if } \max(S^{\#}y_N(t), 2e_N(t)) > bu\}; & \\ \infty, & \\ \qquad\qquad \text{if } \max(S^{\#}y_N(t), 2e_N(t)) \le bu\}; & \end{cases}$$

$$B'_k = \{t \in (0,1) : \eta'(t) < k \le \min(\lambda'(t), \mu'(t))\},$$

$$y'_N = \sum_{k=1}^{N} f_k \kappa_{B'_k}.$$

It is directly checked that

$$B'_k = \{t \in (0,1) : Py_{k-1}(t) > u, \quad Py_{k-1}(t) \le au,$$

$$\max(S^{\#}y_{k-1}(t), 2e_{k-1}(t)) \le bu\} \in \mathcal{F}_{k-1}, \quad k = 1, 2, ..., N.$$

That is why y'_N is a polynomial of the order not greater than N with respect to the H.s. On the set $\{t \in (0,1) : \eta'(t) = \infty\} = \{t \in (0,1) : Py_N(t) \le u\}$ the function Py'_N is equal to zero. Besides, $S^{\#}y'_N \leq 4bu$. Indeed, if $\mu'(t) < \infty$, then

$$\begin{aligned} S^{\#}y'_N(t) &\leq 2S^{\#}y_{\mu'(t)}(t) \leq 2(S^{\#}y_{\mu'(t)-1}(t) + d_{\mu'(t)}\kappa_{A_{\mu'(t)}}(t)) \leq \\ &\leq 2(bu + 2e_{\mu'(t)-1}(t)) \leq 4bu. \end{aligned}$$

If $\mu'(t) = \infty$, then

$$S^{\#}y'_N(t) \leq 2S^{\#}y_N(t) \leq 2bu.$$

Using these facts, we get that

$$\|P_{y'_N}\|_2^2 = \|y'_N\|_2^2 \leq 16b^2u^2 mes\,\{t \in (0,1) : Py_N(t) > u\}. \tag{6.19}$$

Besides, if $Py_N(t) > au$, $\max(S^{\#}y_N(t), 2e_N(t)) \leq bu$, then $\lambda'(t) \leq N$, $\mu'(t) = \infty$, and

$$\begin{aligned} (Py'_N)^2(t) &\geq (Py_{\lambda'(t)})^2(t) - (Py_{\eta'(t)-1})^2(t) - f^2_{\eta'(t)}(t) \geq \\ &\geq a^2u^2 - u^2 - e_N^2(t) \geq (a^2 - 1 - b^2)u^2. \end{aligned}$$

So, we have

$$\{t \in (0,1) : Py_N(t) > au, \quad \max(S^{\#}y_N(t), 2e_N(t)) \leq bu\} \subset$$

$$\subset \{t \in (0,1) : Py'_N(t) > (a^2 - 1 - b^2)^{1/2}u\}.$$

Using the inequality of Khintchine and (6.19), we obtain (6.18):

$$mes\,\{t \in (0,1) : Py_N(t) > au, \max(S^{\#}y_N(t), 2e_N(t)) \leq bu\} \leq$$

$$\leq mes\,\{t \in (0,1) : Py'_N(t) > (a^2-1-b^2)^{1/2}u\} \leq (a^2-1-b^2)^{-1}u^{-2}\|Py'_N\|_2^2 \leq$$

$$\leq 16b^2(a^2 - 1 - b^2)^{-1} mes\,\{t \in (0,1) : Py_N(t) > u\}.$$

By Lemma 6.4, we get that

$$\|Py_N\|_E \leq C_2(E)(\|S^{\#}y_N\|_E + \|e_N\|_E).$$

Inequality (6.17) and the obvious inequalities

$$e_N \leq 2S^{\#}x, \quad Px \leq Py_N + \sum_{k=1}^{N} |\, g_k \,|, \quad S^{\#}y_N \leq S^{\#}x + \sum_{k=1}^{N} |\, g_k \,|$$

allow to complete the proof of the left part of (6.9):

$$\begin{aligned} &\|Px\|_E \leq \|Py_N\|_E + \|\textstyle\sum_{k=1}^N |\, g_k \,|\, \|_E \leq \\ &\leq C_2(E)(\|S^{\#}y_N\|_E + \|e_N\|_E) + 2\|e_N\|_E \leq \\ &\leq C_2(E)(\|S^{\#}x\|_E + \|\textstyle\sum_{k=1}^N |\, g_k \,|\, \|_E + \|S^{\#}x\|_E) \leq C_2(E)\|S^{\#}x\|_E. \end{aligned}$$

□

Corollary 6.5. *Let E be an r.i. space with $\alpha_E > 0$. For every function $x \in E$*

$$c^{-1}(E) \sup_{\varepsilon} \|S^{\#}T_{\varepsilon}x\|_E \leq \|Px\|_E \leq \sqrt{2} \sup_{\varepsilon} \|T_{\varepsilon}x\|_E$$

where the supremum is taken over all choices of signs $\varepsilon = \{\varepsilon_m\}_{m=1}^{\infty}$, $\varepsilon_m = \pm 1$.

Proof. The left inequality follows from (6.9) since for every choice of signs $PT_{\varepsilon}x = Px$. The right inequality follows from the inequality of Khintchine. Indeed,

$$\|Px\|_E \leq \sqrt{2}\| \int_0^1 \sum_{m=1}^{\infty} c_m(x)\chi_m r_m(t)\, dt\|_E \leq$$
$$\leq \sqrt{2}\| \int_0^1 \sum_{m=1}^{\infty} r_m(t) c_m(x)\chi_m\, dt\|_E \leq \sqrt{2} \sup_{\varepsilon} \|T_{\varepsilon}x\|_E.$$

Here $r_m(t) := sign \sin 2^m \pi t$, $m = 1, 2, \ldots;$ $t \in [0, 1]$, are the Rademacher functions.

Theorem 6.6. *The Haar system is an unconditional basis in a separable r.i. space E if and only if $0 < \alpha_E \leq \beta_E < 1$.*

Proof. The "if" part follows from Corollary 4.a.6 and Theorem 6.3. By virtue of these facts, we have

$$C_1(E)\|Px\|_E \leq \|x\|_E \leq C_2(E)\|Px\|_E. \tag{6.20}$$

It is obvious that for every choice of signs $\varepsilon = \{\varepsilon_i\}_{i=1}^{\infty}$, $\varepsilon_i = \pm 1$, $PT_{\varepsilon}x$ is equal to Px. That is why, by Theorem 1.b.5, inequalities (6.20) mean the unconditionality of the H.s. in E.

Necessity. Assume that $\beta_E = 1$. Then, by theorem 1.d.3, for every positive integer n there exist 2^n mutually disjoint functions $\{x_i\}_{i=1}^{2^n}$ all having the same distribution function such that $\|x_i\|_E = 1$ for $1 \leq i \leq 2^n$, and $\{x_i\}_{i=1}^{2^n}$ is 2-equivalent to the unit vector basis of $l_1^{2^n}$. Consider the functions $\{y_i\}_{i=1}^{2^n}$, which are defined by

$$\begin{aligned}
y_1 &:= 2^{-n}(x_1 + x_2 + \ldots + x_{2^n});\\
y_2 &:= 2^{-n}(x_1 + \ldots + x_{2^{n-1}} - x_{2^{n-1}+1} - \ldots - x_{2^n});\\
y_3 &:= 2^{-n}(x_1 + \ldots + x_{2^{n-2}} - x_{2^{n-2}+1} - \ldots - x_{2^{n-1}});\\
y_4 &:= 2^{-n}(x_{2^{n-1}+1} + \ldots + x_{3 \cdot 2^{n-2}} - x_{3 \cdot 2^{n-2}+1} - \ldots - x_{2^n});\\
&\ldots\\
y_{2^{n-1}+1} &:= 2^{-n}(x_1 - x_2);\\
y_{2^n} &:= 2^{-n}(x_{2^n - 1} - x_{2^n}).
\end{aligned}$$

We may assume without loss of generality that each x_i is a finite linear combination of characteristic functions of intervals of the form $(l2^{-k}, (l+1)2^{-k})$ for some fixed k independent of i. Moreover, since all x_i are mutually disjoint and have the same distribution function, it is convenient to assume that, for some integer m and scalars $\{a_j\}_{j=1}^m$, we have

$$x_i = \sum_{j=1}^{m} a_j \kappa_{((i-1+(j-1)2^n)2^{-k},(i+(j-1)2^n)2^{-k})}, \quad 1 \le i \le 2^n.$$

Note that

$$2^n y_2 = \sum_{j=1}^m a_j \chi_{2^{k-n}+j},$$
$$2^n y_3 = \sum_{j=1}^m a_j \chi_{2^{k-n+1}+2j-1},$$
$$2^n y_4 = \sum_{j=1}^m a_j \chi_{2^{k-n}+2j}$$

and so on. In other words, $\{y_i\}_{i=2}^{2^n}$ forms a block basis of the permutation of the H.s. It means that $UC_E(\{y_i\}_{i=2}^{2^n}) \le UC_E(\{\chi_i\}_{i=2}^{\infty})$. On the other hand, the definition of the $\{y_{i=1}\}_{i=2}^{2^n}$ in terms of the $\{x_i\}_{i=1}^{2^n}$ and 2-equivalence of $\{x_i\}_{i=1}^{2^n}$ to the unit vector basis of $l_1^{2^n}$ mean that $UC_E(\{y_i\}_{i=2}^{2^n}) \ge$ $\ge 2^{-1} UC_{L_1}(\{\chi_i\}_{i=1}^{2^n})$. By Remark 5.4, $UC_{L_1}(\{\chi_i\}_{i=1}^{2^n}) \to \infty$ as $n \to \infty$. So, the H.s. is not unconditional in E.

The necessity of the condition $\alpha_E > 0$ is checked in the same manner. □

Another proof of the necessity in Theorem 6.6 is given in the end of Chapter 13 after Remark 13.7.

Chapter 7

Fourier-Haar Coefficients

7.a. The spaces L_p and r.i. spaces

In this section the operator

$$x \to c_{n,k}(x) = c_m(x) = 2^n \int_0^1 x(t)\chi_n^k(t)dt$$

will be investigated. As usual, $m = 2^n + k$, $(n,k) \in \Omega$. Denote

$$d_{n,k}(x) = d_m(x) = 2^{-\frac{n}{2}} c_{n,k}(x).$$

Clearly, $\{d_m(x)\}_1^\infty$ are the Fourier coefficients of $x \in L_1$ with respect to the H.s. normed in L_2. If $x \in L_2$, then, by the Bessel inequality,

$$\lim_{m\to\infty} d_m(x) = 0.$$

In general, it is not true for $x \in L_p$, $1 \le p < 2$. Denote by $L_{2,\infty}^0$ the closure of L_∞ in $L_{2,\infty}$. It was noted in Section 1.c that

$$L_{2,\infty}^0 = \{x : x \in L_{2,\infty},\ \lim_{\tau\to 0} \tau^{-\frac{1}{2}} \int_0^\tau x^*(t)\,dt = 0\}.$$

The analogy of the Merser theorem is

Theorem 7.a.1. *Let E be an r.i. space. Then, for each $x \in E$,*

$$\lim_{m\to\infty} d_m(x) = 0 \tag{7.a.1}$$

iff

$$E \subset L_{2,\infty}^0. \tag{7.a.2}$$

Proof. Necessity. If imbedding (7.a.2) is not valid, then there exists a function $x = x^* \in E$ such that

$$\limsup_{\tau\to 0}\tau^{-\frac{1}{2}}\int_0^\tau x(t)\,dt \geq 1.$$

We can find a sequence of integers $\{m_i\}_1^\infty$, such that

$$\int_0^{2^{-m_i}} x(t)\,dt \geq 2^{-\frac{m_i+1}{2}}, \quad 1 = 1,2,\dots \ .$$

We may assume that

$$x(2^{-m_{i+1}}) \geq 2\,x(2^{-m_i}), \quad i = 1,2,\dots \ .$$

Let us construct the function

$$y(t) = \begin{cases} \min\left(\frac{1}{2}x(2^{-m_{i+1}}), x(t)\right), & t \in (2^{-m_{i+1}}, 2^{-m_i}], \\ x(t), & t \in (2^{-m_1}, 1]. \end{cases}$$

Clearly, for every $t \in (0,1]$,

$$\frac{1}{2}x(t) \leq y(t) \leq x(t). \tag{7.a.3}$$

If, for some $i \in \{2,3,\dots\}$, $t < 2^{-m_i} < s$, then

$$y(t) \geq 2y(s). \tag{7.a.4}$$

By (7.a.3,4),

$$\begin{aligned} d_{m_i-1,1}(y) = \ & 2^{\frac{m_i-1}{2}}\int_0^1 y(t)\chi^1_{m_i-1}(t)\,dt = 2^{\frac{m_i-1}{2}}\Big(\int_0^{2^{-m_i}} y(t)\,dt - \\ & - \int_{2^{-m_i}}^{2^{-m_i+1}} y(t)\,dt\Big) \geq 2^{\frac{m_i-1}{2}-1}\int_0^{2^{-m_i}} y(t)\,dt \geq \\ & \geq 2^{\frac{m_i-1}{2}-2}\int_0^{2^{-m_i}} x(t)\,dt \geq 2^{\frac{m_i-1}{2}-2-\frac{m_i+1}{2}} = \tfrac{1}{8}. \end{aligned}$$

This proves the necessity of condition (7.a.2).

Sufficiency. If $x \in L^0_{2,\infty}$, then

$$\lim_{\tau\to 0}\tau^{-1/2}\int_0^\tau x^*(t)\,dt = 0. \tag{7.a.5}$$

By (1.c.1),

$$\Big|\int_0^1 x(t)\chi^k_n(t)\,dt\Big| \leq \int_{(k-1)2^{-n}}^{k2^{-n}} |\,x(t)\,|\,dt \leq \int_0^{2^{-n}} x^*(t)\,dt.$$

Therefore

$$| d_{n,k}(x) |= 2^{\frac{n}{2}} \left| \int_0^1 x(t)\chi_n^k(t)\,dt \right| \le 2^{\frac{n}{2}} \int_0^{2^{-n}} x^*(t)\,dt,$$

and (7.a.5) implies

$$\lim_{m\to\infty} d_m(x) = 0. \quad \square$$

A more general statement is true.

Theorem 7.a.2. *Let E be an r.i. space and $1 < p < \infty$. Then, for every $x \in E$,*

$$\lim_{m\to\infty} m^{\frac{1}{p}} c_m(x) = 0$$

iff $E \subset L^0_{p,\infty}$.

The proof is completely analogous to the given above.

Theorem 7.a.2 may be formulated in a different form. Let us consider the linear operator

$$A_p x = \{m^{-\frac{1}{p}} c_m(x)\}_1^\infty.$$

Then the operator A_p is bounded from an r.i. space E into $c_0(l_\infty)$ iff E is imbedded in $L^0_{p,\infty}(L_{p,\infty})$. Let us mention that the condition

$$\sup_m m^{-\frac{1}{p}} \mid c_m(x) \mid < \infty$$

does not imply $x \in L_{p,\infty}$. Indeed, let

$$x_j(t) = \sum_{n=1}^{j} \sum_{k=1}^{2^n} \chi_n^k(t), \quad j = 1, 2, \ldots$$

Then

$$c_{n,k}(x_j) = \begin{cases} 1, & 1 \le n \le j; \\ 0, & n = 0, \quad n > j; \end{cases}$$

and

$$\sup_{m\ge 1} m^{-\frac{1}{p}} c_m(x_j) = 3^{-\frac{1}{p}}.$$

On the other hand, by (1.d.7) and the Khintchines inequality

$$\|x_j\|_{L_{p,\infty}} \ge \|x_j\|_{L_1} \ge \sqrt{\frac{j}{2}}.$$

Consequently, the inequality

$$\|A_p x\|_{l_\infty} \ge a\|x\|_{L_{p,\infty}}$$

is impossible for any $a > 0$.

However, this problem is more simple for monotone functions. At first we note that the Fourier-Haar coefficients of any decreasing function are non-negative.

Theorem 7.a.3. *Let x be a non-increasing and non-negative function on [0,1], $x \in L_1$, $1 < p < \infty$. Then $x \in L_{p,\infty}$ iff*

$$\sup_{n \in \mathbf{N}} 2^{-\frac{n}{p}} c_{n,1}(x) < \infty.$$

Proof. The necessity was established in Theorems 7.a.1 and 7.a.2. To prove the sufficiensy we consider the function

$$\phi(t) = \int_0^t x(s)\, ds.$$

This function is positive and concave on [0,1], $\phi(0) = 0$. Denote

$$\phi_n = \phi(2^{-n}) = \int_0^{2^{-n}} x(s)\, ds.$$

We may assume that $\phi_0 = 1$ and

$$\sup_{n \in \mathbf{N}} 2^{-\frac{n}{p}} c_{n,1}(x) \leq 1.$$

Then

$$\phi_{n+1} - (\phi_n - \phi_{n+1}) = \int_0^1 x(s)\chi_n^1(s)\, ds = 2^{-n} c_{n,1}(x) \leq 2^{\frac{n}{p}-n}$$

Hence,

$$\phi_{n+1} \leq 2^{\frac{n}{p}-n-1} + \frac{\phi_n}{2}.$$

Iterating this inequality, we get

$$\begin{aligned}\phi_{n+1} &\leq 2^{\frac{n}{p}-n-1} + 2^{\frac{n-1}{p}-n-1} + \tfrac{\phi_{n-1}}{4} \leq \textstyle\sum_{k=0}^n 2^{\frac{n-k}{p}-n-1} + \tfrac{\phi_0}{2^{n+1}} \leq \\ &\leq \frac{2^{\frac{n}{p}-n-1}}{1-2^{-\frac{1}{p}}} + 2^{-n-1} \leq \frac{2^{\frac{n}{p}-n}}{1-2^{-\frac{1}{p}}}.\end{aligned}$$

So,

$$\phi_n 2^{n-\frac{n}{p}} \leq \frac{2^{1-\frac{1}{p}}}{1-2^{-\frac{1}{p}}}, \qquad n = 1, 2, \ldots \quad .$$

It means that

$$\|x\|_{L_{p,\infty}} = \sup_{0<t\leq 1} t^{\frac{1}{p}-1}\phi(t) \leq 2^{\frac{1}{p}} < \sup_{n \in \mathbf{N}} 2^{-\frac{n}{p}+n}\phi_n \leq \frac{2}{1-2^{-\frac{1}{p}}}. \qquad \Box$$

An analogous statement is true for the spaces L_p. The norm of a non-negative and decreasing function depends only on its first Fourier-Haar coefficients up to equivalence.

Lemma 7.a.4. *Let $1 < p < \infty$. There exists a constant $C = C(p) > 0$ such that, for every sequence $\{a_k\}$, the following inequalities*

$$\frac{1}{C}\left(\sum_{n=0}^{\infty} | a_n |^p 2^{-n}\right)^{\frac{1}{p}} \leq \| \sum_{n=0}^{\infty} a_n \chi_n^1 \|_{L_p} \leq C\left(\sum_{n=0}^{\infty} | a_n |^p 2^{-n}\right)^{\frac{1}{p}}$$

hold.

Proof. By Theorem 1.b.8 and Corollary 5.8, there is a constant $C = C(p)$ such that

$$\| \sum_{n=0}^{\infty} a_n \chi_n^1 \|_{L_p} \geq \tfrac{1}{C}\left(\int_0^1 (\sum_{n=0}^{\infty} (a_n \chi_n^1(t))^2)^{\frac{p}{2}}\, dt\right)^{\frac{1}{p}} \geq$$

$$\geq \tfrac{1}{C}\left(\int_0^1 \sum_{n=0}^{\infty} | a_n |^p \,\kappa_{(0,2^{-n})}(t)\, dt\right)^{\frac{1}{p}} \geq \tfrac{1}{C}(\sum_{n=0}^{\infty} | a_n |^p 2^{-n})^{\frac{1}{p}}.$$

Using the obtained inequality and the Holder inequality we have

$$\| \sum_{n=0}^{\infty} a_n \chi_n^1 \|_{L_p} = \sup_{\{b_n\}\neq 0} \frac{\int_0^1 (\sum_{n=0}^{\infty} a_n \chi_n^1(t))(\sum_{n=0}^{\infty} b_n \chi_n^1(t))\, dt}{\| \sum_{n=0}^{\infty} b_n \chi_n^1 \|_{L_{p'}}} \leq$$

$$\leq \sup_{\{b_n\}\neq 0} \frac{C \sum_{n=0}^{\infty} a_n b_n 2^{-n}}{(\sum_{n=0}^{\infty} |b_n|^{p'} 2^{-n})^{1/p'}} = C(\sum_{n=0}^{\infty} | a_n |^p 2^{-n})^{1/p},$$

where $\frac{1}{p} + \frac{1}{p'} = 1$. □

Lemma 7.a.4 may be formulated in the equivalent form. The system $\{\chi_n^1\}_{n=0}^{\infty}$ normed in L_p is equivalent to the standart basis of l_p, $1 < p < \infty$. Clearly, this statement fails for $p = 1$ and $p = \infty$.

Theorem 7.a.5. *Let $1 < p < \infty$, $x(t)$ be a non-negative and non-increasing function on [0,1]. Then*

$$\|x\|_{L_p} \approx c_{00}(x) + (\sum_{n=0}^{\infty} c_{n,1}^p(x) 2^{-n})^{\frac{1}{p}}.$$

Proof. Given a finite sequence $a_n \geq 0$, we have

$$\sum_{n=0}^{\infty} a_n c_{n,1}(x) 2^{-n} = \sum_{n=0}^{\infty} a_n \int_0^1 x(s)\chi_n^1(s)\, ds =$$

$$= \int_0^1 (\sum_{n=0}^{\infty} a_n \chi_n^1(s))\, x(s)\, ds \leq \| \sum_{n=0}^{\infty} a_n \chi_n^1 \|_{L_{p'}} \|x\|_{L_p}.$$

By Lemma 7.a.4,

$$\| \sum_{n=0}^{\infty} a_n \chi_n^1 \|_{L_{p'}} \leq C(p') \left(\sum_{n=0}^{\infty} a_n^{p'} 2^{-n}\right)^{\frac{1}{p'}}$$

Hence,

$$\sum_{n=0}^{\infty} a_n c_{n,1}(x) \leq C(p')(\sum_{n=0}^{\infty} a_n^{p'} 2^{-n})^{\frac{1}{p'}} \|x\|_{L_p}$$

and

$$(\sum_{n=0}^{\infty} c_{n,1}^p(x) 2^{-n})^{\frac{1}{p}} = \sup_{(\sum_{n=0}^{\infty} a_n^{p'} 2^{-n})^{\frac{1}{p'}} \leq 1} \sum_{n=0}^{\infty} a_n c_{n,1}(x) \leq C(p')\|x\|_{L_p}.$$

So, the first part of the Theorem is proved.

Given a function $u = u^*$, we construct the function

$$v(t) = \sum_{n=1}^{\infty} u(2^{-n}) \kappa_{(2^{-n}, 2^{-n+1}]}(t).$$

Clearly, $v \in Q := [\chi_0^0, \ \chi_n^1, \ n \geq 0]$ and

$$u(t) \leq v(t) \leq u(\frac{t}{2}).$$

Consequently,

$$\|x\|_{L_p} \leq 2 \sup_{y \in Q, \|y\|_{L_{p'}} \leq 1} \int_0^1 x(t) y(t)\, dt.$$

There exists a function $y \in Q \cap L_{p'}$ such that $\|y\|_{L_{p'} \leq 1}$ and

$$\int_0^1 x(t) y(t)\, dt \geq \frac{1}{3} \|x\|_{L_p}.$$

Clearly, Lemma 7.a.4 is valid for each $y \in Q$. There is a constant $C = C(p) > 0$ and numbers $\{a_n\}$, $a_n \geq 0$, $n \geq -1$ such that

$$y(t) = a_{-1} \chi_0^0(t) + \sum_{n=0}^{\infty} a_n \chi_n^1(t)$$

and

$$(a_{-1}^p + \sum_{n=0}^{\infty} a_n^p 2^{-n})^{\frac{1}{p}} \leq C.$$

Applying the Holder inequality, we obtain

$$(c_{0,0}^p(x) + \sum_{n=0}^{\infty} c_{n,1}^p(x) 2^{-n})^{1/p} \geq \frac{1}{C}(a_{-1} c_{0,0}(x) + \sum_{n=0}^{\infty} 2^{-n} a_n c_{n,1}(x)) =$$

$$= \frac{1}{C} \int_0^1 (a_{-1} \chi_0^0(t) + \sum_{n=0}^{\infty} a_n \chi_n^1(t)) x(t)\, dt \geq \frac{1}{3C} \|x\|_{L_p}. \quad \square$$

Simple examples and also Theorem 7.a.2 show that, in general, for $x \in L_1$, the sequence $c_m(x)$ does not tend to 0. Denote

$$\sigma_j(x) = \frac{1}{j} \sum_{m=1}^{j} c_m(x), \quad j = 1, 2, \ldots$$

Theorem 7.a.6. *For every $x \in L_1$,*

$$\lim_{j\to\infty} \sigma_j(x) = 0. \tag{7.a.6}$$

Proof. Given $j > 2$, we find an integer i such that $2^i \leq j < 2^{i+1}$. Then

$$\begin{aligned} \sigma_j(x) &= \tfrac{1}{j}\big(c_{00}(x) + \textstyle\sum_{n=0}^{i-1} \sum_{k=1}^{2^n} c_{n,k}(x) + \sum_{k=1}^{j-2^i} c_{i,k}(x)\big) = \\ &= \tfrac{1}{j} \textstyle\int_0^1 x(t)\big(1 + \sum_{n=0}^{i-1} 2^n r_{n+1}(t) + 2^i \sum_{k=1}^{j-2^i} \chi_i^k(t)\big)\, dt, \end{aligned} \tag{7.a.7}$$

where $r_n(t) = \operatorname{sign} \sin 2^n \pi t$ are the Rademacher functions. Since

$$\frac{1}{j} \mid 1 + \sum_{n=0}^{i-1} 2^n r_{n+1}(t) + 2^i \sum_{k=1}^{j-2^i} \chi_i^k(t) \mid \leq \frac{1}{j}(1 + \sum_{n=0}^{i} 2^n) = \frac{2^{i+1}}{j} \leq 2,$$

then

$$\mid \sigma_j(x) \mid \leq 2 \|x\|_{L_1}. \tag{7.a.8}$$

Clearly, $\lim_{j\to\infty} \sigma_j(x) = 0$ for each polynomial x with respect to the H.s. The set of polynomials is dence in L_1. Therefore (7.a.8) implies (7.a.6). □

If $x \in L_p$, $p > 1$, then (7.a.6) may be strengthened.

Theorem 7.a.7. *Let $p > 1$ and $x \in L_p$. There exists a constant $C = C(p)$ such that*

$$\|\{\sigma_{2^i}(x)\}\|_{l_2} \leq C \|x\|_{L_p}.$$

Proof. First we note that the operator

$$Ay = \{\sum_{n=1}^{i} 2^{n-i} y_n\}_{i=1}^{\infty}$$

is bounded in l_p, $1 \leq p \leq \infty$. This simple statement is well known. We present the proof for the completeness of the exposition. If $\mid y_n \mid \leq 1$ for every $n = 1, 2, \ldots$, then

$$\mid (Ay)_i \mid \leq \sum_{n=1}^{i} 2^{n-i} < 2.$$

Consequently, $\|A\|_{l_\infty} \leq 2$. If $y \in l_1$, then

$$\|Ay\|_{l_1} = \sum_{i=1}^{\infty} | \sum_{n=1}^{\infty} 2^{n-i} y_n | \leq \sum_{n=1}^{\infty}\sum_{i=n}^{\infty} 2^{n-i} | y_n | = 2\|y\|_{l_1}.$$

By the Riesz interpolation theorem (1.e.2)

$$\|A\|_{l_p} \leq 2$$

for every $p \in [1, \infty]$.

Applying (7.a.7), we get

$$\sigma_{2^i} x = \int_0^1 x(t) \left(2^{-i} + \sum_{n=0}^{i-1} 2^{n-i} r_{n+1}(t)\right) dt =$$

$$2^{-i} \int_0^1 x(t)\, dt + \sum_{n=0}^{i-1} 2^{n-i} b_{n+1}(x),$$

where $\{b_n(x)\}$ are the Fourier coefficients of x with respect to the Rademacher system $\{r_n\}$. Put

$$Bx = \{b_n(x)\}_1^\infty, \quad Dx = \{2^{-i} \int_0^1 x(t)\, dt\}_1^\infty.$$

Then $\sigma = AB + D$, $\sigma x = \{\sigma_{2^i} x\}_1^\infty$. It is evident that D is bounded from L_1 into l_1 and, consequently, from L_p into l_2. The subspace $[r_n]$ is complemented in L_p([163], 2.b.4). Therefore the Khintchine's inequality implies that B is bounded from L_p into l_2. □

The last statement may be generalized. It was proved in [240] that the operator B is bounded from $L_{M(\frac{1}{2})}$ into l_2, where $M(\alpha) = M(\alpha, u) = |u| \ln^\alpha(1+|u|)$, $\alpha > 0$. Therefore σ is bounded from $L_{M(\frac{1}{2})}$ into l_2. By Theorem 7.a.7, the operator σ acts from L_1 into c_0. Applying the complex interpolation method to the pairs $(L_1, L_{M(\frac{1}{2})})$, (c_0, l_2) and using the known equalities

$$[L_1, L_{M(\frac{1}{2})}]_\Theta = L_{M(\frac{\Theta}{2})}, \quad [c_0, l_2]_\Theta = l_{\frac{2}{\Theta}}$$

[35], we get that σ acts from $L_{M(\alpha)}$ into $l_{\frac{1}{\alpha}}$, $0 < \alpha < \frac{1}{2}$.

On the other hand, Theorem 7.a.7 fails if $p = 1$. To construct a counterexample we consider the sequence of functions $x_m(t) = 2^m \kappa_{(0,2^{-m})}(t)$, $m = 1, 2, \ldots$. We have

$$x_m(t) = \chi_0^0(t) + \sum_{n=0}^{m-1} 2^n \chi_n^1(t).$$

If $1 \leq j \leq m$, then

$$\sigma_{2^j}(x_m) = 2^{-j}(1 + \sum_{n=0}^{j-1} 2^n c_{n,1}(x_m)) = 2^{-j}(1 + \sum_{n=0}^{j-1} 2^n) = 1.$$

Hence,

$$\|\{\sigma_{2^j}(x_m)\}\|_{l_2} \geq m^{\frac{1}{2}}.$$

Consequantly, σ is not bounded from L_1 into l_2.

Let E be a Banach lattice and $\{\chi_n^k\}$ forms an unconditional basis in E. By Theorem 1.b.8,

$$\|\sum_{(n,k) in \Omega} c_{n,k}\chi_n^k\|_E \approx \| \left(\sum_{(n,k)\in\Omega} c_{n,k}^2 \kappa_{\Delta_n^k} \right)^{1/2} \|_E. \qquad (7.a.9)$$

For L_p $(1 < p < \infty)$, this formula may be given in a more precise form. The assumption

$$\sum_{(n,k)\in\Omega} c_{n,k}\chi_n^k \in L_p$$

implies that

$$\lim_{N\to\infty} \| \sum_{(n,k)\in\Omega, n\leq N} c_{n,k}\chi_n^k\|_{L_p} = \| \sum_{(n,k)\in\Omega} c_{n,k}\chi_n^k\|_{L_p}.$$

Therefore,

$$\|\sum\nolimits_{(n,k)\in\Omega} c_{n,k}\chi_n^k\|_{L_p} \approx | c_{00} | + \\ + \lim\nolimits_{N\to\infty} \left(2^{-N} \sum\nolimits_{\{k_n\}\in T_N} (\sum\nolimits_{n=0}^N c_{n,k_n}^2)^{\frac{p}{2}} \right)^{\frac{1}{p}}, \qquad (7.a.10)$$

where T_N is the set of sequences $\{k_0, k_1, ..., k_N\}$ such that $k_0 = 1$ and $2k_{n-1} - k_n = 0$ or 1 for every $n = 1, 2, ..., N$. Clearly, $| T_N |= 2^N$. The equivalence constants depend only on $p \in (1, \infty)$.

It is known that the inequality

$$\left(\sum_{n=0}^{\infty} 2^{-n} (\sum_{k=1}^{2^n} | c_{n,k} |^p)^{\frac{2}{p}} \right)^{\frac{1}{p}} \leq M_p \| \sum_{(n,k)\in\Omega} c_{n,k}\chi_n^k\|_{L_p}$$

holds for $p \in (1, 2]$. This is valid for a more wide set of spaces.

Theorem 7.a.8. *Let E be a separable r.i.space and $\beta_E < 1$. The estimate*

$$(\sum_{n=0}^{\infty} \| \sum_{k=1}^{2^n} a_{n,k}\chi_n^k\|_E^2)^{1/2} \leq M \| \sum_{(n,k)\in\Omega} a_{n,k}\chi_n^k\|_E \qquad (7.a.11)$$

holds for some constant $M > 0$ and every finite sequence $\{a_{n,k}\}$, $(n, k) \in \Omega$, iff E is 2-concave.

Proof. Necessity. Let us prove that the lower Boyd index of E is positive. Let n, m be integers, $n > m$ and

$$x_1(t) = \sum_{k=1}^{2^m} a_k \chi_n^k(t).$$

Put

$$x_2(t) = \sum_{k=1}^{2^m} a_k (\chi_{n+1}^{2^{m+1}+2k-1}(t) + \chi_{n+1}^{2^{m+1}+2k}(t)) = B_n x_1(t).$$

By analogy we put

$$x_{j+1} = B_{n+j-1} x_j, \quad 1 \le j \le 2^{n-m}.$$

Clearly, the functions x_j are equimeasurable and disjoint, $1 \le j \le 2^{n-m}$. Assumption (7.a.11) implies

$$(2^{n-m} \|x_1\|_E^2)^{1/2} \le M \|\sigma_{2^{n-m}} x_1\|_E.$$

Hence, for every $n = 1, 2, ...,$

$$\|\sigma_{2^n}\| \ge \frac{1}{M} 2^{\frac{n}{2}}.$$

It means that $\alpha_E \ge \frac{1}{2}$. So, $\frac{1}{2} \le \alpha_E \le \beta_E < 1$.

By Theorem 6.6, the H.s. forms an unconditional basis in E. Hence, by (7.a.9), there exists a constant $M_1 > 0$ such that

$$\frac{1}{M_1} \| \left(\sum_{(n,k)\in\Omega} a_{n,k}^2 \kappa_{\Delta_n^k} \right)^{1/2} \|_E \le \| \sum_{(n,k)\in\Omega} a_{n,k} \chi_n^k \|_E \le$$

$$\le M_1 \| \left(\sum_{(n,k)\in\Omega} a_{n,k}^2 \kappa_{\Delta_n^k} \right)^{1/2} \|_E \tag{7.a.12}$$

for all numbers $\{a_{n,k}\}$. Uniting (7.a.11) and the right part of (7.a.12), we get

$$\left(\sum_{n=0}^{\infty} \| \sum_{k=1}^{2^n} a_{n,k} \chi_n^k \|_E^2 \right)^{1/2} \le M M_1 \| (\sum_{(n,k)\in\Omega} a_{n,k}^2 \kappa_{\Delta_n^k})^{1/2} \|_E.$$

Denote

$$\sum_{k=1}^{2^n} a_{n,k} \kappa_n^k(t) = x_n(t).$$

Then the last inequality may be written as follows

$$(\sum_{n=0}^{\infty} \|x_n\|_E^2)^{1/2} \le M M_1 \| (\sum_{n=0}^{\infty} x_n^2)^{1/2} \|_E. \tag{7.a.13}$$

The left and right parts of (7.a.13) do not depend on the sign of $x_n(t)$. Therefore (7.a.13) is true for every function

$$x_n(t) = \sum_{k=1}^{2^n} a_{n,k}\kappa_{\Delta_n^k(t)}.$$

The set of such functions is dense in E. Consequently, (7.a.13) is valid for each $x_n \in E$.

Sufficiency. If E is 2-concave, then the inequality

$$(\sum_{n=0}^{N} \|\sum_{k=1}^{2^n} a_{n,k}\chi_n^k\|^2)^{1/2} \le M_2\|(\sum_{n=0}^{N}\sum_{k=1}^{2^n} a_{n,k}^2\kappa_{\Delta_n^k})^{1/2}\|_E$$

is true for some constant M_2, for every integer N and for every sequence $\{a_{n,k}\}$. It was noted in Section 1.d that 2-concavity of E implies $\alpha_E > \frac{1}{2}$. If $0 < \alpha_E \le \beta_E < 1$, then equivalence (7.a.9) takes place. Hence,

$$(\sum_{n=0}^{\infty} \|\sum_{k=1}^{2^n} a_{n,k}\chi_n^k\|_E^2)^{1/2} \le M_1M_2\|\sum_{(n,k)\in\Omega} a_{n,k}\chi_n^k\|_E. \quad \square$$

Given $s \in (1,\infty)$, we consider the quasi-linear operator

$$V_s x(t) = \sum_{n=1}^{\infty}(\sum_{k=1}^{s} | c_{n,k} |^s)^{1/s}\kappa_{(2^{-n},2^{-n+1})}(t)$$

defined on L_1.

Lemma 7.a.9. *If $1 \le r \le s$, then the operator V_s has the weak type (r,r).*

Proof. Let $x \in L_r$. Put

$$D = \{n : n \ge 1, \quad \sum_{k=1}^{2^n} | c_{n,k} |^s \ge 1\}$$

and

$$m = \min\{n : n \in D\}.$$

Then

$$mes\,\{t : V_s x(t) \ge 1\} = \sum_{n\in D} 2^{-n} \le 2^{-m+1}.$$

Applying Theorem 4.a.1, we get

$$\int_0^1 | \sum_{k=1}^{2^m} c_{m,k}\chi_m^k(t) |^r \, dt \le \int_0^1 | \sum_{n=m}^{\infty}\sum_{k=1}^{2^n} c_{n,k}\chi_n^k(t) |^r \, dt \le$$

$$\le 2^r \int_0^1 | \sum_{n=0}^{\infty}\sum_{k=1}^{2^n} c_{n,k}\chi_n^k(t) |^r \, dt.$$

Since

$$\int_0^1 \Big| \sum_{k=1}^{2^m} c_{m,k}\chi_m^k(t) \Big|^r = 2^{-m} \sum_{k=1}^{2^m} | c_{m,k} |^r \geq 2^{-m} \left(\sum_{k=1}^{2^m} | c_{m,k} |^s \right)^{\frac{r}{s}} \geq 2^{-m},$$

then

$$mes\,\{t : V_s x(t) \geq 1\} \leq 2^{r+1} \int_0^1 \Big| \sum_{n=0}^{\infty} \sum_{k=1}^{2^n} c_{n,k}\chi_n^k(t) \Big|^r \, dt.$$

Hence, for any $\tau > 0$,

$$\tau (mes\{t : V_s x(t) \geq \tau\})^{\frac{1}{r}} \leq 2^{1+\frac{1}{r}} \|x\|_{L_r} \leq 4 \|x\|_{L_r}. \quad \square$$

Applying the Marcinkiewicz interpolation theorem (1.e.1), we have

Corollary 7.a.10. *Let $1 < p < s < \infty$. There exists a constant $M(p,s)$ such that*

$$\left(\sum_{n=0}^{\infty} 2^{-n} \Big(\sum_{k=1}^{2^n} | c_{n,k}(x) |^s \Big)^{\frac{p}{s}} \right)^{\frac{1}{p}} \leq M(p,s) \|x\|_{L_p}. \qquad (7.a.14)$$

This Corollary may be strengthened.

Theorem 7.a.11. *Let E be an r.i. space, $s > 1$. If $\frac{1}{s} < \alpha_E \leq \beta_E < 1$, then there exists a constant $M = M(s,E)$ such that*

$$\|V_s x\|_E \leq M \|x\|_E$$

for each $x \in E$.

To prove this theorem it is sufficient to apply Theorem 1.e.2 to the operator V_s.

Note that the operator V_s is not bounded in L_1 and L_s, $s < 2$. Indeed, if

$$x_N(t) = \sum_{n=1}^{N} 2^n \chi_n^1(t), \quad N = 1, 2, \ldots,$$

then $\|x_N\|_{L_1} < 2$ and

$$\|V_s x_N\|_{L_1} = \Big\| \sum_{n=1}^{N} 2^n \kappa_{(2^{-n}, 2^{-n+1})} \Big\|_{L_1} = N.$$

Consequently, V_s is not bounded in L_1 for any s.

Put

$$y_N(t) = \sum_{n=1}^{N} \sum_{k=1}^{2^n} \chi_n^k(t) = \sum_{n=1}^{N} r_{n+1}(t), \quad N = 1, 2, \ldots \ .$$

The Rademacher functions form an orthonormal system. For $s < 2$, we have

$$\|y_N\|_{L_s} \leq \|y_N\|_{L_2} = N^{1/2}$$

and

$$V_s y_N(t) = \sum_{n=1}^{N} 2^{\frac{n}{s}} \kappa_{(2^{-n},2^{-n+1})}(t).$$

Hence,

$$\|V_s y_N\|_{L_s} = \left(\int_0^1 \sum_{n=1}^{N} 2^n \kappa_{(2^{-n},2^{-n+1})}(t)\, dt\right)^{\frac{1}{s}} = N^{\frac{1}{s}}.$$

Therefore the operator V_s is not bounded in L_s, $s < 2$. This statement fails for $s \geq 2$.

Theorem 7.a.12. *1. If* $1 \leq p \leq 2$, *then*

$$\| \sum_{(n,k)\in\Omega} a_{n,k}\chi_n^k\|_{L_p} \leq \left(\sum_{(n,k)\in\Omega} | a_{n,k} |^p \, 2^{-n} \right)^{\frac{1}{p}} \tag{7.a.15}$$

2. If $2 \leq p \leq \infty$, *then*

$$\left(\sum_{(n,k)\in\Omega} | a_{n,k} |^p \, 2^{-n} \right)^{\frac{1}{p}} \leq \| \sum_{(n,k)\in\Omega} a_{n,k}\chi_n^k\|_E. \tag{7.a.16}$$

Proof. We consider the space $l_p(\Omega)$ equipped with the norm

$$\|a\|_{l_p(\Omega)} = \left(\sum_{(n,k)\in\Omega} | a_{n,k} |^p \, 2^{-n} \right)^{1/p}$$

and the operator

$$Ta = \sum_{(n,k)\in\Omega} a_{n,k}\chi_n^k(t).$$

By the Parseval equality, T acts isometrically from $l_2(\Omega)$ into L_2. The obvious inequality

$$\| \sum_{(n,k)\in\Omega} a_{n,k}\chi_n^k\|_{L_1} \leq \sum_{(n,k)\in\Omega} | a_{n,k} | \, 2^{-n}$$

shows that

$$\|Ta\|_{L_1} \leq \|a\|_{l_1(\Omega)}.$$

By the Riesz interpolation theorem, T is bounded from $l_p(\Omega)$ into L_p and

$$\|Ta\|_{L_p} \leq \|a\|_{l_p(\Omega)}$$

for every $a \in l_p(\Omega)$, $p \in [1,2]$. Inequalities (7.a.15) and (7.a.16) are dual. □

7.b. Absolutely continuous functions

The Fourier-Haar coefficients of an absolutely continuous function may be expressed via its derivative. We introduce the seminorm

$$\|x\|_W = \int_0^1 | x'(t) | \, dt$$

in the space $W = W_1^1$ of absolutely continuous functions. Let us consider the sequence of operators

$$T_n x(t) = 2^{n+2} \sum_{k=1}^{2^n} c_{n,k}(x) \kappa_{\Delta_n^k(t)}.$$

For $(n,k) \in \Omega_1$, we put

$$\phi_{n,k}(t) = \rho(t, [0,1] \backslash \Delta_n^k) = \begin{cases} 0, & t \in \bar{\Delta}_n^k; \\ t - (k-1)2^{-n}, & t \in \bar{\Delta}_{n+1}^{2k-1}; \\ k2^{-n} - t, & t \in \bar{\Delta}_{n+1}^{2k}. \end{cases}$$

It is easy to check that

$$\phi_{n,k}(t) = \max\left(\min((k - \frac{1}{2})2^{-n}, t) - \max(k-1)2^{-n}, t - 2^{-n-1}), 0 \right)$$

If $x \in W$, $\quad (n,k) \in \Omega_1$, then

$$c_{n,k}(x) = 2^n \int_{(k-1)2^{-n}}^{(2k-1)2^{-n-1}} x(t)\,dt - 2^n \int_{(2k-1)2^{-n-1}}^{k2^{-n}} x(t)\,dt =$$

$$= 2^n \int_{(k-1)2^{-n}}^{(2k-1)2^{-n-1}} (x(t) - x(t + 2^{-n-1}))\,dt =$$

$$= -2^n \int_{(k-1)2^{-n}}^{(2k-1)2^{-n-1}} \int_t^{t+2^{-n-1}} x'(s)\,ds =$$

$$= -2^n \int_{(k-1)2^{-n}}^{k2^{-n}} x'(s) \int_{\max((k-1)2^{-n}, s-2^{-n-1})}^{\min((2k-1)2^{-n-1}, s)} dt\,ds =$$

$$= -2^n \int_0^1 x'(s) \phi_{n,k}(s)\,ds.$$

Hence,

$$T_n x(t) = -\int_0^1 K_n(t,s) x'(s)\,ds,$$

where

$$K_n(t,s) = 2^{2n+2}\sum_{k=1}^{2^n}\phi_{n,k}(s)\kappa_{\Delta_n^k(t)}.$$

Theorem 7.b.1. *Let E be a separable r.i. space, $x \in W$ and $x' \in E$. Then*

$$\lim_{n\to\infty}\|T_n x + x'\|_E = 0.$$

Proof. Since

$$\int_{(k-1)2^{-n}}^{k2^{-n}} \phi_{n,k}(s)\,ds = 2^{-2n-2}, \quad (n,k) \in \Omega_1,$$

then

$$\int_0^1 K_n(t,s)\,ds = 1 \quad \text{a.e. } t \in [0,1].$$

It means that the operators

$$Q_n x(t) = \int_0^1 K_n(t,s)x(s)\,ds$$

are uniformly bounded in L_∞. By analogy with this statement, Q_n are uniformly bounded in L_1. By Theorem 1.e.4, Q_n are uniformly bounded in E and $\|Q_n\|_E = 1$.

Given $x \in W_1^1$ and an integer m, denote by $x_m(t)$ a function such that $x_m(k2^{-m}) = x(k2^{-m})$ for each $k = 0,1,\dots,2^m$ and $x_m(t)$ is linear on $\bar\Delta_m^k$ for each $k = 1,2,\dots,2^m$. If $x(t) = \alpha t + \beta$ for every t belonging to some integral $\bar\Delta_n^k$, then

$$2^{n+2}c_{n,k}(x) \equiv -\alpha. \tag{7.b.1}$$

It means that

$$T_n x_m(t) = -\frac{d}{dt}x_m(t),$$

where $n \geq m$, $t \neq k2^{-m}$, $k = 1,2,\dots,2^m-1$. Clearly,

$$\frac{d}{dt}x_m = S_{2^m}x'.$$

Recall that $S_{2^m}x$ is the partial sum of x. By Theorem 4.a.3,

$$\lim_{m\to\infty}\|x' - x_m'\|_E = 0.$$

Therefore the uniform boundedness of Q_n in E implies that

$$\lim_{n\to\infty}\|T_n x + x'\|_E = 0. \quad \square$$

Corollary 7.b.2. *Let* $1 \le p < \infty$, $x \in W_1^p$. *Then*

$$\lim_{n\to\infty} 2^{p(n+2)-n} \sum_{k=1}^{2^n} | c_{n,k}(x) |^p = \int_0^1 | x'(t) |^p \, dt. \tag{7.b.2}$$

Denote by V the space of functions on [0,1] having the bounded variation. The seminorm

$$\|x\|_V = \sup \sum_{i=1}^{n} | x(t_i) - x(t_{i-1}) |$$

is introduced in V, where the supremum is taken on the integers $n = 1, 2, ...$ and the increasing sequences $0 \le t_0 < t_1 < ... < t_n \le 1$.

Theorem 7.b.3. *If* $x \in V$, *then*

$$\sum_{k=1}^{2^n} | c_{n,k}(x) | \le \frac{1}{2}\|x\|_V.$$

Proof. Let $n \in \{0, 1, ...\}$. We can find a sequence $\{t_i\}$, $1 \le i \le 2^{n+1}$ such that $t_i \in \Delta_{n+1}^i$ and

$$\left| \int_{\Delta_{n+1}^{2k-1}} x(t)\, dt - \int_{\Delta_{n+1}^{2k}} x(t)\, dt \right| \le 2^{-n-1} | x(t_{2k-1}) - x(t_{2k}) |$$

for every $k = 1, 2, ..., 2^n$. Hence,

$$\sum_{k=1}^{2^n} | c_{n,k}(x) | = 2^n \sum_{k=1}^{2^n} \left| \int_{\Delta_{n+1}^{2k-1}} x(t)\, dt - \int_{\Delta_{n+1}^{2k}} x(t)\, dt \right| \le$$

$$\le \tfrac{1}{2} \sum_{k=1}^{2^n} | x(t_{2k-1}) - x(t_{2k}) | \le \tfrac{1}{2}\|x\|_V.$$

□

Theorem 7.b.4. *Let* $x \in W_1^1$. *Then*

$$\lim_{n\to\infty} \sum_{k=1}^{2^n} c_{n,k}(x) = \frac{1}{4}(x(0) - x(1)).$$

Proof. We consider the functionals

$$f_n(x) = \sum_{k=1}^{2^n} c_{n,k}(x), \quad n = 0, 1, ...$$

on W_1^1. By Theorem 7.b.3,

$$| f_n(x) | \le \frac{1}{2}\|x\|_V = \frac{1}{2}\|x\|_W.$$

Consequently,

$$\sup_n \|f_n\| \le \frac{1}{2}. \tag{7.b.3}$$

Denote by Q the set of piecewise linear functions for which the break points are contained in the set of dyadic numbers. The Fourier-Haar coefficients of a piecewise linear function were checked in (7.b.1). Hence, for $x \in Q$ and sufficiently large n,

$$f_n(x) = \frac{1}{4}(x(0) - x(1)). \tag{7.b.4}$$

Since Q is dence in W_1^1, then (7.b.3) and (7.b.4) imply

$$\lim_{n\to\infty} f_n(x) = \frac{1}{4}(x(0) - x(1))$$

for every $x \in W_1^1$. □

Note that Theorem 7.b.4 fails for continuous functions. Repeating the proof of Theorem 7.b.4, we can obtain a more general result. If $0 \le \alpha < \beta \le 1$ and $x \in W_1^1$, then

$$\lim_{n\to\infty} \sum_{\alpha<k2^{-n}<\beta} c_{n,k}(x) = \frac{1}{4}(x(\alpha) - x(\beta)) = -\frac{1}{4}\int_\alpha^\beta x'(t)\,dt. \tag{7.b.5}$$

Hence,

$$\lim_{n\to\infty} \sum_{k=1}^{2^n} f(k2^{-n})c_{n,k}(x) = -\frac{1}{4}\int_0^1 f(t)x'(t)\,dt \tag{7.b.6}$$

for any $x \in W_1^1$ and any step function f. By Theorem 7.b.3,

$$\sup_n 2^n \sum_{k=1}^{2^n} | c_{n,k}(x) | \le \frac{1}{2}\int_0^1 | x'(t) | \,dt.$$

Using this estimate, it is easy to extend (7.b.9) from the set of step functions on the set of continuous functions. So, we have proved

Theorem 7.b.5. *Let $x \in W_1^1$, $f \in C$. Then*

$$\lim_{n\to\infty} \sum_{k=1}^{2^n} f(k2^{-n})c_{n,k}(x) = -\frac{1}{4}\int_0^1 f(t)x'(t)\,dt.$$

7.c. Continuous functions

The H.s. discontinuity implies that the Fourier-Haar coefficients of a continuous function can not tend to 0 very quickly.

Theorem 7.c.1. *Let $x \in C$, $t_0, t_1 \in [0,1]$, $t_0 < t_1$. Then*

$$\limsup_{2^n+k\to\infty} \mid c_{n,k}(x) \mid 2^n \geq \frac{1}{4} \mid \frac{x(t_0)-x(t_1)}{t_0-t_1} \mid . \tag{7.c.1}$$

Proof. Without loss of generality we may assume that $c_{00} = 0$. At first we prove that

$$\limsup_{2^n+k\to\infty} \mid c_{n,k} \mid 2^n = \sup_{(n,k)\in\Omega} \mid c_{n,k} \mid 2^n. \tag{7.c.2}$$

Supposing the contrary, we can find an element $(m,j) \in \Omega_1$ such that

$$\mid c_{n,k} \mid 2^n < \mid c_{m,j} \mid 2^m$$

for every $(n,k) \in \Omega$, $n > m$. The function

$$\sum_{n=1}^{2^m+j-1} c_n \chi_n(t)$$

is continuous in the point $t_0 = (2j-1)2^{-m-1}$. The jump of $c_{mj}\chi_m^j(t)$ in t_0 is equal to $2 \mid c_{mj} \mid$. The jump of

$$\sum_{n=2^m+j+1}^{\infty} c_n \chi_n(t)$$

in t_0 does not exceed

$$2 \sum_{n=m+1}^{\infty} \max_{1\leq k\leq 2^n} \mid c_{n,k} \mid < 2 \sum_{n=m+1}^{\infty} \mid c_{mj} \mid 2^{m-n} = 2 \mid c_{mj} \mid,$$

Therefore $x(t)$ must be discontinuous in the point t_0. It proves (7.c.2).

Given $t_0, t_1 \in [0,1]$, $t_0 \neq t_1$, there exist $s_0, s_1 \in [t_0, t_1]$ and a dyadic interval Δ_p^q such that $\mid x(t_0) - x(t_1) \mid \leq 2 \mid x(s_0) - x(s_1) \mid$ and $s_0, s_1 \in \Delta_p^q$, $s_1 - s_0 \geq 2^{-p-1}$. Let

$$\sup_{(n,k)\in\Omega} \mid c_{n,k} \mid 2^n = \limsup_{2^n+k} \mid c_{n,k} \mid 2^n = a.$$

Then $\mid c_{n,k}\chi_n^k(t) \mid \leq a2^{-n}$ for every $(n,k) \in \Omega$. Hence,

$$\mid x(t_0) - x(t_1) \mid \leq 2 \mid x(s_0) - x(s_1) \mid \leq \sum_{n=p}^{\infty} a2^{-n} = a2^{-p+1} \leq$$

$$\leq 4a(s_1 - s_0) \leq 4a(t_1 - t_0).$$

The obtained inequality implies (7.c.1). □

Note that the lim sup in (7.c.2) may be taken over only $(n,k) \in \Omega$ such that $\Delta_n^k \subset [t_0, t_1]$. Theorem 7.c.1 states that if $x \in C$ and

$$\lim_{n\to\infty} n\, c_n(x) = 0, \tag{7.c.3}$$

then x is a constant on [0,1].

Theorem 7.c.2. *Let $x \in C$ and*

$$\sum_{m=1}^{\infty} | c_m(x) | < \infty. \tag{7.c.4}$$

Then x=const.

Proof. For each integer n, we have

$$\|c_n \chi_n\|_V \leq 4 \mid c_n \mid$$

Therefore the series

$$\sum_{n=1}^{\infty} c_n \chi_n$$

is absolutely convergent in V. By Theorem 3.1,

$$\sum_{n=1}^{\infty} c_n \chi_n = x.$$

Consequently,

$$\lim_{m \to \infty} \|x - \sum_{n=1}^{m} c_n \chi_n\|_V = 0. \tag{7.c.5}$$

On the other hand, x is continuous and

$$\sum_{n=1}^{m} c_n \chi_n$$

is a step function. Hence,

$$\|x - \sum_{n=1}^{m} c_n \chi_n\|_V = \|x\|_V + \|\sum_{n=1}^{m} c_n \chi_n\|_V \geq \|x\|_V. \tag{7.c.6}$$

It follows from (7.c.5) and (7.c.6) that $\|x\|_V = 0$. □

In fact, if

$$\sum_{(n,k) \in \Omega} | a_{n,k} | < \infty$$

and

$$| a_{n,k} | \geq q \max(| a_{n+1,2k-1} |, | a_{n+1,2k} |)$$

take place for some $q > 2$ and for every $(n, k) \in \Omega$, then the function

$$\sum_{(n,k) \in \Omega} a_{n,k} \chi_n^k(t)$$

is continuous in a point $t \in (0,1)$ iff t is dyadically irrational.

Given $\alpha \in (0,1]$, we consider the space $Lip\,\alpha$ equipped with the seminorm

$$\|x\|_{Lip\,\alpha} = \sup_{t_1 \neq t_2} \frac{|\, x(t_1) - x(t_2) \,|}{|\, t_1 - t_2 \,|^{\alpha}}.$$

Theorem 7.c.3. *Let $x \in C$. Then $x \in Lip\,\alpha$ iff*

$$\sup_{(n,k)\in\Omega_1} 2^{n\alpha} \mid c_{n,k}(x) \mid < \infty.$$

Proof. Let

$$\mid x(t_1) - x(t_2) \mid \leq c \mid t_1 - t_2 \mid^{\alpha}$$

for some constant $c > 0$ and for every $t_1, t_2 \in [0,1]$. Let $(n,k) \in \Omega_1$. Using (2.3), we get

$$\mid c_{n,k}(x) \mid = 2^n \mid \textstyle\int_{\Delta_n^k} (x(t) - x(t + 2^{-n-1}))\, dt \mid \leq$$

$$\leq 2^n \textstyle\int_{\Delta_n^k} \mid x(t) - x(t + 2^{-n-1}) \mid dt \leq 2^n c (2^{-n-1})^{\alpha} 2^{-n-1} = c 2^{-n\alpha - \alpha - 1}.$$

Hence,

$$\sup_{(n,k)\in\Omega_1} 2^{n\alpha} \mid c_{n,k}(x) \mid \leq 2^{-\alpha-1} \|x\|_{Lip\,\alpha}.$$

Suppose that $\mid c_{n,k}(x) \mid \leq a 2^{-n\alpha}$ for some $a > 0$ and every $(n,k) \in \Omega_1$, $m \geq 1$. Then

$$\mid \sum_{(n,k)\in\Omega, n\geq m} c_{n,k} \chi_n^k(t) \mid \leq \sum_{n=m}^{\infty} a 2^{-n\alpha} = \frac{a 2^{-m\alpha}}{1 - 2^{-\alpha}}.$$

It means that

$$\|x - S_{2^m} x\|_{L_\infty} \leq \frac{a 2^{-m\alpha}}{1 - 2^{-\alpha}}.$$

By (3.2),

$$\omega(2^{-m}, x) \leq \frac{4a 2^{-m\alpha}}{1 - 2^{-\alpha}}.$$

For $\tau \in [0,1]$, we have

$$\omega(\tau, x) \leq \frac{4a(2\tau)^{\alpha}}{1 - 2^{-\alpha}}.$$

Therefore

$$\|x\|_{Lip\alpha} \leq \frac{2^{2+\alpha}}{1 - 2^{-\alpha}} \sup_{(n,k)\in\Omega_1} 2^{n\alpha} \mid c_{n,k}(x) \mid.$$

□

7.d. Characteristic functions

We consider the functional

$$H(e) = \sup_{(n,k)\in\Omega_1} | c_{n,k}(\kappa_e) |$$

on the σ-algebra of measurable sets of [0,1]. Clearly, $0 \le H(e) \le \frac{1}{2}$ for each $e \subset [0,1]$. If $0 < mes\, e < 1$, then $H(e) > 0$. Put $h(s) = H(\kappa_{(0,s)})$. Since

$$\inf_{2^{-n-1}\le s\le 2^{-n}} h(s) = \inf_{2^{-n}\le s\le 2^{-n+1}} h(s), \quad n = 1, 2, ...,$$

then

$$\inf_{0<s<1} h(s) = \inf_{\frac{1}{4}\le s\le \frac{1}{2}} h(s) = \inf_{\frac{1}{4}\le s\le \frac{1}{2}} \max(s, 1-2s) = \frac{1}{3}.$$

Consequently, for $s \in (0,1)$, we have

$$\frac{1}{3} \le h(s) \le \frac{1}{2}.$$

Both inequalities are exact. Indeed, it is easy to check that $h(\frac{1}{3}) = \frac{1}{3}$ and $h(s) = \frac{1}{2}$ if s is dyadically rational. The inequality $H(e) \ge \frac{1}{3}$ is valid for every $e \subset [0,1]$ such that e is a finite union of some intervals. In general it fails for every measurable set of [0,1].

Theorem 7.d.1. *Given an integer N, there exists a measurable set $e \subset [0,1]$ such that $mes\, e = \frac{1}{2}$ and $H(e) = \frac{1}{2N}$.*

Proof. We define the numbers $\varepsilon_{n,k} \in \{0,1\}$, $(n,k) \in \Omega$ by induction. Put $\varepsilon_{00} = 0$, $\varepsilon_{0,1} = 1$ and suppose that $\varepsilon_{n,k}$ are defined for every $(n,k) \in \Omega$, $n < m$. Denote

$$x_m(t) = \sum_{(n,k)\in\Omega_{n<m}} \varepsilon_{n,k}\chi_n^k(t)$$

and put

$$\varepsilon_{mj} = \begin{cases} 1, & \text{if} \quad | x_m(t) | < N \quad \text{for} \quad t \in \Delta_m^j; \\ 0, & \text{if} \quad | x(t) | = N \quad \text{for} \quad t \in \Delta_m^j. \end{cases}$$

Denote by Q the set of $t \in [0,1]$ such that the series

$$\sum_{(n,k)\in\Omega} \varepsilon_{n,k}\chi_n^k(t) \tag{7.d.1}$$

is divergent.

Let us prove that $mes\, Q = 0$. If $t_0 \in Q$, then $\chi_n^{k_n}(t) = \pm 1$ for some $k_n \in \{1, 2, ..., 2^{n-1}\}$ and $\varepsilon_{n,k_n} = 1$ for every $n = 0, 1, ...$ Since

$$| \sum_{k=r}^{r+2N} \chi_n^{k_n}(t_0) | \le 2N, r = 0, 1, ..,$$

then

$$-1, 1 \in \{\chi_n^{k_n}(t_0), r \le n \le r + 2N\}.$$

Every $t \in [0, 1]$ may be written in the form

$$t = \sum_{n=1}^{\infty} i_n 2^{-n}$$

where $i_n = i_n(t) = 0$ or 1. If $t_0 \in Q$, then

$$0, 1 \in \{i_n(t_0), \quad r \le n \le r + 2N\}, \quad r = 0, 1, 2, \ldots \quad . \tag{7.d.2}$$

It is well known that the set of numbers satisfying condition (7.d.2) for some N has the zero measure. Therefore $mes\, Q = 0$. So, we have proved that series (7.d.1) converges a.e.

Clearly,

$$mes\,\{t : \sum_{(n,k)\in\Omega} \varepsilon_{n,k}\chi_n^k(t) = N\} = mes\{t : \sum_{(n,k)\in\Omega} \varepsilon_{n,k}\chi_n^k(t) = -N\} = \frac{1}{2}.$$

Denote

$$e = \{t : \sum_{(n,k)\in\Omega} \varepsilon_{n,k}\chi_n^k(t) = N\}.$$

Then

$$c_{n,k}(\kappa_e) \in \{0, \frac{1}{2N}\}.$$

for any $(n, k) \in \Omega$. Hence,

$$H(e) = \frac{1}{2N}.$$

□

Chapter 8

The Haar system and martingales

Definition 8.1. *Let $\beta_1 \subset \beta_2 \subset ... \subset \beta_n \subset ...$ be an increasing sequence of σ-subalgebras of Σ. A sequence $\{x_n\}_{n=1}^{\infty}$ of integrable functions is said to be a martingale with respect to $\{\beta_n\}_{n=1}^{\infty}$ if*

$$E_{\beta_n} x_{n+1} = x_n,$$

for all n.

Put $d_1 = x_1, \quad d_n = x_n - x_{n-1}, \quad n = 2, 3, ...$ It is clear that a sequence $\{x_n\}_{n=1}^{\infty}$ is a martingale if and only if d_n is β_n-measurable, for every n, and

$$E_{\beta_n} d_{n+1} = 0,$$

for all n. A sequence $\{d_n\}_{n=1}^{\infty}$ with such properties is called a martingale - difference sequence.

Lemma 8.2. *A martingale-difference sequence $\{d_n\}_{n=1}^{\infty}$ is a basic one in every r.i. space E.*

Proof. By theorem 1.e.6, for every $n = 1, 2, ...$ and every sequence of scalars $\{c_k\}_{k=1}^{n+1}$, we have

$$\|\sum_{k=1}^{n} c_k d_k\|_E = \|E_{\beta_n}(\sum_{k=1}^{n+1} c_k d_k)\|_E \le \|\sum_{k=1}^{n+1} c_k d_k\|_E. \tag{8.1}$$

This means that $\{d_n\}_{n=1}^{\infty}$ is a basic sequence in E (see Theorem 1.b.1). □

By equations (2.2), the H.s. is a martingale-difference sequence with respect to $\{\mathcal{F}_m\}_{m=1}^{\infty}$. The same is valid for every block basis of the H.s. Indeed, let

$$d_k = \sum_{m=m_{k-1}+1}^{m_k} c_m \chi_m,$$

where, $0 = m_0 < m_1 < m_2 < ...,\quad k = 1, 2, ...$ Then, using (2.2) and the properties of $E_{\mathcal{F}}$ (see Section 1.a), we get

$$E_{\mathcal{F}_{m_k}} d_{k+1} = \sum_{m=m_k+1}^{m_{k+1}} c_m E_{\mathcal{F}_{m_k}} \chi_m = \sum_{m=m_k+1}^{m_{k+1}} c_m E_{\mathcal{F}_{m-1}} \chi_m = 0.$$

The following theorem shows that, up to equivalence, the converse statement is valid.

Theorem 8.3. *Let E be an r.i. space, $\{d_n\}_{n=1}^{\infty}$ be a martingale-difference sequence of functions from E. Then, for every $\varepsilon > 0$, there exists a block basis of the H.s. which is $(1+\varepsilon)$-equivalent to $\{d_n\}_{n=1}^{\infty}$.*

Proof. Since the set of step-functions is dense in E, and the basic sequences possess stability properties (Theorem 1.b.2), it sufficies to prove Theorem 8.3 only for the case when all the functions d_n, $n = 1, 2, ...$, are step-functions, and the correspoding σ-subalgebras β_n are generated by finite decompositions of $[0, 1]$. Remark now that there always exists a sequence of σ-algebras $\{\mathcal{A}_k\}_{k=1}^{\infty}$ such that:

a) $\beta_n = \mathcal{A}_{k_n}$, $n = 1, 2, \dots$, for some sequence $k_1 < k_2 < \dots$;

b) for every $k = 1, 2, ...$ $\mathcal{A}_{k+1}$ is obtained from $\mathcal{A}_k$ by decomposing one atom of $\mathcal{A}_k$ into two subsets of non-zero measure.

In other words, $\mathcal{A}_1 \subset \mathcal{A}_2 \subset ... \subset \mathcal{A}_k \subset \mathcal{A}_{k+1} \subset ...$, and $\mathcal{A}_k$ is generated by decomposing $[0, 1]$ into k intervals, $k = 1, 2, ...$

If we put $e_1 = E_{\mathcal{A}_1} d_1$ and

$$e_k = E_{\mathcal{A}_k} d_n - E_{\mathcal{A}_{k-1}} d_n,$$

for $k \geq 2$, $k \in \{k_{n-1}+1, k_{n-1}+2, ..., k_n\}$, $n = 1, 2...$, $k_0 := 0$, then the sequence $\{e_k\}_{k=1}^{\infty}$ will be martingale-difference, and, for $n = 1, 2, ...$

$$d_n = \sum_{k=k_{n-1}+1}^{k_n} e_k.$$

So, it suffices to prove Theorem 8.3 for $\{e_k\}_{k=1}^{\infty}$. Without loss of generality we may assume that, besides properties a) and b), σ-subalgebras $\mathcal{A}_k$ satisfy the following: if $\mathcal{A}_{k+1}$ is obtained from $\mathcal{A}_k$ by decomposing the atom A of $\mathcal{A}_k$ into two disjoint subset A_1 and A_2, then the ratio $mes\,(A_1)/mes\,(A)$ is equal to $l \cdot 2^{-m}$ for some positive integers l and m.

For such σ-subalgebras $\{\mathcal{A}_k\}_{k=1}^{\infty}$ it is possible to construct by induction σ-subalgebras $\{\mathcal{A}'_k\}_{k=1}^{\infty}$ with the following properties:

a) there exists one-to-one correspondence between the atoms of $\mathcal{A}_k$ and those of $\mathcal{A}'_k$, and the measures of corresponding atoms are equal;

b) $\mathcal{A}'_k \subset \mathcal{F}_{n_k}$ for some sequence $1 = n_1 < n_2 < ...$;

c) $E_{\mathcal{A}'_k} x = E_{\mathcal{F}_{n_k}} x$ for every $\mathcal{A}'_{k+1}$ - measurable function x.

Let us describe one step of induction. Let the σ-subalgebras $\mathcal{A}'_1, ..., \mathcal{A}'_k$ be defined. Choose n_k such that $\mathcal{A}'_k \subset \mathcal{F}_{n_k}$. Let the σ-subalgebra $\mathcal{A}_{k+1}$ be obtained from $\mathcal{A}_k$ by decomposing the atom A of $\mathcal{A}_k$ into two disjoint subsets A_1 and A_2 , where $mes\,(A_1)/mes\,(A_2) = l \cdot 2^{-m}$. Note that the atom A', corresponding to A, is the union of disjoint dyadic intervals $D_1, ..., D_p$ for some positive integer p, where every D_i is an atom of $\mathcal{F}_{n_k}$. If $D_i = (j2^{-n}, (j+1)2^{-n})$, then put $D'_i = (j2^{-n}, j2^{-n} + l2^{-n-m})$, $i = 1, 2, ..., p$. Define

$$A'_1 = \cup_{i=1}^{p} D'_i, \quad A'_2 = A' - A'_1.$$

It is obvious that

$$\frac{mes\,(A'_1)}{mes\,(A)} = l2^{-m}, \quad mes\,(A'_1) = mes\,(A_1), \quad mes\,(A'_2) = mes\,(A_2).$$

Besides,

$$E_{\mathcal{F}_{n_k}} \kappa_{A'_1} = l2^{-m} \kappa_{A'} = E_{\mathcal{A}'_k} \kappa_{A'_1},$$

$$E_{\mathcal{F}_{n_k}} \kappa_{A'_2} = (1 - l2^{-m}) \kappa_{A'} = E_{\mathcal{A}'_k} \kappa_{A'_2}.$$

Define $\mathcal{A}'_{k+1}$ as a σ-subalgebra generated by $\mathcal{A}'_k$, A'_1 and A'_2. By virtue of the last two equations, property c) is valid.

Using property a) it is possible to construct the martingale-difference sequence $\{e'_k\}_{k=1}^{\infty}$ with respect to $\{\mathcal{A}'_k\}_{k=1}^{\infty}$ which is 1-equivalent to $\{e_k\}_{k=1}^{\infty}$. Since $E_{\mathcal{F}_{n_k}} e'_{k+1} = E_{\mathcal{A}'_k} e'_{k+1} = 0$, then

$$e'_{k+1} = \sum_{j=n_k+1}^{n_{k+1}} c_j \chi_j, \quad k = 1, 2, ... \quad \Box$$

Corollary 8.4. *Let E be a separable r.i. space, and let $\{d_n\}_{n=1}^{\infty}$ be a martingale-difference sequence of functions from E. Then we have*

$$C_E(\{d_n\}_{n=1}^{\infty}) \le C_E(\{\chi_m\}_{m=1}^{\infty}).$$

If $0 < \alpha_E \le \beta_E < 1$, then $\{d_n\}_{n=1}^{\infty}$ is an unconditional basic sequence in E, and

$$UC_E(\{d_n\}_{n=1}^{\infty}) \le UC_E(\{\chi_m\}_{m=1}^{\infty}). \tag{8.2}$$

For L_p, $1 < p < \infty$, in particular, we have, for every martingal-difference sequence,

$$UC_{L_p}(\{d_n\}_{n=1}^{\infty}) \le p^* - 1. \tag{8.3}$$

Remark 8.5. Now we shall prove that the constant $p^* - 1$ is the best possible one in Theorem 5.5. Let $p \ge 2$ and fix some positive number a. Define $w > p$ by the equation

$$a^p + pw^{p-1} - w^p = 0.$$

Denote by $\Theta = 1 - 1/w$, and put, for $k = 1, 2, \ldots$, $\beta_k = 1 - w\delta/(a + k\delta)$, where δ is some number from the interval $(0, a/w)$. Consider the sequence of functions

$$d_1 := a\kappa_{[0,1)}; \quad d_k = \delta\kappa_{[0,\prod_{i=1}^{k-1}\beta_i)} +$$

$$+[\Theta(a + (k-1)\delta) - (a + (k-2)\delta)]\,\kappa_{[\prod_{i=1}^{k-1}\beta_i,\prod_{i=1}^{k-2}\beta_i)}, \quad k = 2, 3, \ldots .$$

Since $\int_0^1 d_k\,dt = 0$ for $k = 2, 3, \ldots$, and the functions d_l with $l < k$ are constant on $supp\, d_k$, then the sequence $\{d_m\}_{m=1}^{\infty}$ is a martingal-difference one.

Not complicated calculations show that

$$\lim_{a\to 0}\lim_{\delta\to 0}\lim_{n\to\infty} \| \textstyle\sum_{k=1}^{n}(-1)^k d_k\|_{L_p} = 1,$$

$$\lim_{a\to 0}\lim_{\delta\to 0}\lim_{n\to\infty} \| \textstyle\sum_{k=1}^{n} d_k\|_{L_p} = p - 1.$$

These equations together with inequality (8.2) prove that the constant $p-1$ is the best possible one in Theorem 5.5 if $p \geq 2$. For $p \in (1, 2)$ consider the functions

$$d'_k = (q-1)d_k,\ k = 1, 2, \ldots;\ q := p/(p-1).$$

Then

$$\lim_{a\to 0}\lim_{\delta\to 0}\lim_{n\to\infty} \| \textstyle\sum_{k=1}^{n}(-1)^k d'_k\|_{L_p} = q - 1,$$

$$\lim_{a\to 0}\lim_{\delta\to 0}\lim_{n\to\infty} \| \textstyle\sum_{k=1}^{n} d'_k\|_{L_p} = (p-1)(q-1) = 1.$$

If $\{x_n\}_{n=1}^{\infty}$ is an unconditional basic sequence in a Banach lattice E, then, generally speaking, it is not true that, for every sequence of reals $\{a_n\}_{n=1}^{\infty}$,

$$\|(\sum_{n=1}^{\infty} a_n^2 x_n^2)^{1/2}\|_E \approx \|\sum_{n=1}^{\infty} a_n x_n\|_E$$

(see, for example, [163, c. 51]). However, for martingale-difference sequences in r.i. spaces E with non-trivial Boyd indices the last equivalence is valid. This follows from the given below two theorems.

Let us introduce some notations:

$$\bar{x} = \{x_n\}_{n=1}^{\infty},$$

$$d_1 = x_1, \quad d_k = x_k - x_{k-1}, \quad k = 2, 3, \ldots;$$

$$P\bar{x} = (\textstyle\sum_{k=1}^{\infty} d_k^2)^{1/2}, \quad x_n^{\#} = \sup_{1\leq k\leq n} | x_k |,$$

$$\bar{x}^{\#} = \sup_{1\leq n<\infty} | x_n |, \quad \|\bar{x}\|_E = \sup_{1\leq n<\infty} \|x_n\|_E.$$

Theorem 8.6. *Let E an r.i. space. The following assertions are equivalent:*

1) There exists a constant C such that for every martingale $\bar{x} = \{x_n\}_{n=1}^{\infty}$ we have

$$\|\bar{x}\|_E \leq \|\bar{x}^{\#}\|_E \leq C\|\bar{x}\|_E; \tag{8.4}$$

2) $\lim_{\tau\to\infty} \|\sigma_\tau\|_E/\tau = 0.$

To prove the theorem we need

Lemma 8.7. *Let $\bar{x}$ be a martingale, $1 \leq p < \infty$. Then, for every $u > 0$ and for every $n \in N$,*

$$u^p mes\,\{t \in (0,1) : x_n^{\#}(t) > u\} \leq \int_{\{t\in(0,1):x_n^{\#}(t)>u\}} | x_n(s) |^p \; ds \leq \|x_n\|_{L_p}^p .$$

The proof of the lemma is similar to that of Theorem 4.a.4.

Proof of Theorem 8.6. We shall prove first that 2) implies 1). The left inequality in (8.4) is obvious. By lemmas 8.7 and 4.a.7, we get that for every $n \in \mathbf{N}$

$$\|x_n^{\#}\|_E \leq C\|x_n\|_E \leq C\|\bar{x}\|_E.$$

Since an r.i. space has the Fatou property (see Section 1.d), the last inequality implies the right part of (8.4).

To prove that 1) implies 2) we assume the contrary and use the sequences $\{x_i\}_{i=1}^{2^n}$ and $\{y_i\}_{i=1}^{2^n}$ defined in Theorem 6.3. It is clear that $\{y_i\}_{i=1}^{2^n}$ is martingale difference, and we have the following equalities:

$$y_1 + y_2 + \textstyle\sum_{k=1}^{n-1} 2^k y_{2^k+1} = x_1,$$

$$\sup_{1\leq k\leq n-1} | \, y_1 + y_2 + \textstyle\sum_{l=1}^{k} 2^l y_{2^l+1} \,| = | \, x_1 + 2^{-1}x_2 + 2^{-2}(x_3 + x_4) +$$

$$+ 2^{-3}(x_5 + x_6 + x_7 + x_8) + \ldots + 2^{2-n}(x_{2^{n-3}+1} + \ldots + x_{2^{n-2}}) \,| \, .$$

Since $\{x_i\}_{i=1}^{2^n}$ is 2-equivalent in E to the unit vector basis of $l_1^{2^n}$, we have

$$\|y_1 + y_2 + \textstyle\sum_{k=1}^{n-1} 2^k y_{2^k+1}\|_E \leq 2,$$

$$\| \sup_{1\leq k\leq n-1} | \, y_1 + y_2 + \textstyle\sum_{l=1}^{k} 2^l y_{2^l+1} \,| \, \|_E \geq 2^{-1} + 2^{-2}(n-2).$$

These inequalities contradict (8.4). □

Theorem 8.8. *Let E be an r.i. space. The following assertions are equivalent:*

1) There exist some constants C_1 and C_2 such that for every martingale $\bar{x}$ we have

$$C_1\|P\bar{x}\|_E \leq \|\bar{x}^{\#}\|_E \leq C_2\|P\bar{x}\|_E. \tag{8.5}$$

2) $\lim_{s\to 0} \|\sigma_s\|_E = 0.$

To prove the theorem we need two lemmas.

Lemma 8.9. *Let $(S, \mathcal{F}, \mu)$ be a σ-finite measure space. Let $\mathcal{F}_0 \subset \mathcal{F}_1 \subset \mathcal{F}_2 \subset \dots$ be an increasing sequence of σ-subalgebras of $\mathcal{F}$, and let $\{z_i\}_{i=1}^{\infty}$ be a sequence of non-negative $\mathcal{F}$-measurable functions. Put $W = \sum_{i=1}^{\infty} E_{\mathcal{F}_{i-1}} z_i$, $V = \sum_{i=1}^{\infty} z_i$. Then for every positive λ we have*

$$\int_{\{t: W(t) > \lambda\}} (W(t) - \lambda)\, d\mu \le \int_{\{t: W(t) > \lambda\}} V(t)\, d\mu. \tag{8.6}$$

Proof. Put

$$W_n = \sum_{i=1}^{n} E_{\mathcal{F}_{i-1}} z_i, \quad V_n = \sum_{i=1}^{n} z_i, \quad n \in N.$$

It is obvious that $W_n \uparrow W$ and $V_n \uparrow V$ if $n \to \infty$. So, it suffices to prove (8.6) for W_n and V_n, where $n \in \mathbf{N}$. Note that

$$\{t : W_n(t) > \lambda\} = \cup_{i=1}^{n} A_i, \tag{8.7}$$

where $A_i = \{t : W_{i-1}(t) \le \lambda,\ W_i(t) > \lambda\}$, and the sets $\{A_i\}_{i=1}^{n}$ are mulually disjoint.

Let us prove that, for $i = 1, 2, \dots, n$, we have

$$\int_{A_i} W_n(t)\, d\mu \le \lambda \mu(A_i) + \int_{A_i} V_n(t)\, d\mu. \tag{8.8}$$

Indeed,

$$\int_{A_i} W_n(t)\, d\mu = \int_{A_i} W_{i-1}(t)\, d\mu + \int_{A_i} (W_n(t) - W_{i-1}(t))\, d\mu \le$$

$$\le \lambda \mu(A_i) + \int_{A_i} (W_n(t) - W_{i-1}(t))\, d\mu. \tag{8.9}$$

Since $A_i \in \mathcal{F}_{i-1}$, then

$$\int_{A_i} (W_n(t) - W_{i-1}(t))\, d\mu = \int_{A_i} \sum_{j=i}^{n} E_{\mathcal{F}_{j-1}} z_j(t)\, d\mu \le$$

$$\le \int_{A_i} \sum_{j=i}^{n} z_j(t)\, d\mu \le \int_{A_i} V_n(t)\, d\mu. \tag{8.10}$$

Inequality (8.8) follows from (8.9) and (8.10). Summing up inequalities (8.8) with respect to $i = 1, \dots, n$ and taking into account (8.7), we get (8.6) for W_n and V_n, $n \in \mathbf{N}$. □.

Lemma 8.10. *Let E be an r.i. space and the $\lim_{\tau\to 0}\|\sigma_\tau\|_E = 0$. If two non-negative functions x and y from E satisfy inequality*

$$\int_{\{t:x(t)>\lambda\}} x(s)\,ds \le \lambda\, mes\,\{t : x(t) > \lambda\} + \int_{\{t:x(t)>\lambda\}} y(s)\,ds, \qquad (8.11)$$

for every $\lambda > 0$, then $\|x\|_E \le C\|y\|_E$, where the constant C depends only on E.

Proof. It follows from (8.11) that

$$\|x\|_{L_1} \le \|y\|_{L_1} \le \|y\|_E. \qquad (8.12)$$

By the reasoning similar to that used in Lemma 4.a.7 we get from (8.11) that, for $t \in [0,1]$,

$$\int_0^1 x^*(s)\,ds \le tx^*(t) + \int_0^t y^*(s)\,ds. \qquad (8.13)$$

Fix $\tau_0 \in (0,1)$ such that $\|\sigma_{\tau_0}\|_E < 1$. Note that, for $t \in [0,1]$,

$$\int_0^t x^*(s)\,ds \ge \int_0^{\tau_0 t} x^*(s)\,ds + (1-\tau_0)tx^*(t) =$$

$$= \tau_0 \int_0^t x^*(\tau_0 s)\,ds + (1-\tau_0)tx^*(t) = \tau_0 \int_0^t \sigma_{1/\tau_0}x^*(s)\,ds + (1-\tau_0)tx^*(t).$$

The last inequality together with (8.13) implies that, for every $t \in (0,1)$, we have

$$\int_0^t \sigma_{1/\tau_0}x^*(s)\,ds \le tx^*(t) + \frac{1}{\tau_0}\int_0^t y^*(s)\,ds \le \int_0^t \left(x^*(s) + \frac{1}{\tau_0}y^*(s)\right)\,ds.$$

Using Theorem 1.e.5, we get

$$\|\sigma_{1/\tau_0}x^*\|_E \le \|x\|_E + \frac{1}{\tau_0}\|y\|_E. \qquad (8.14)$$

It is clear that

$$\sigma_{\tau_0}\sigma_{1/\tau_0}x^* = x^*\kappa_{(0,\tau_0)},$$

where κ_e is characteristic function of set $e \subset [0,1]$. That is why,

$$\|x\|_E \le \|\sigma_{\tau_0}\|_E \cdot \|\sigma_{1/\tau_0}x^*\|_E + x^*(\tau_0). \qquad (8.15)$$

It follows from (8.14) and (8.15) that

$$(1 - \|\sigma_{\tau_0}\|_E)\|x\|_E \le \frac{\|\sigma_{\tau_0}\|_E}{\tau_0}\|y\|_E + x^*(\tau_0).$$

Using the Chebyshev inequality and (8.12), we have

$$x^*(\tau_0) \leq \frac{\|x\|_{L_1}}{\tau_0} \leq \frac{\|y\|_E}{\tau_0}.$$

So, we finally obtain that

$$\|x\|_E \leq \frac{1+\|\sigma_{\tau_0}\|_E}{\tau_0(1-\|\sigma_{\tau_0}\|_E)}\|y\|_E. \quad \square$$

Proof of Theorem 8.8. Sufficiency of the condition $\lim_{s\to 0}\|\sigma_s\|_E = 0$ is proved according to the scheme used in theorem 6.3. However, the reasoning is more complicated because the arbitrary martingale-difference sequence does not possess the specific property of the H.s. :

$$|\chi_k| = \kappa_{\Delta_k}, \quad \Delta_k \in \mathcal{F}_{k-1}, \quad k = 2, 3, \dots .$$

So, let $\bar{x} = \{x_i\}_{i=1}^{\infty}$ be a martingale with respect to $\mathcal{A}_1, \mathcal{A}_2, \dots$; let $\bar{d} = \{d_i\}_{i=1}^{\infty}$ be its martingale-difference sequence.

Put

$$d_0 = e_0 = 0, \; e_k = \max_{1\leq i\leq k} |d_i|, \; e = \sup_{1\leq k<\infty} |d_k|,$$

$$A_k = \{t \in (0,1) : |d_k(t)| \leq 2e_{k-1}(t)\},$$

$$f_k = d_k\kappa_{A_k}, \quad g_k = d_k - f_k,$$

$$a_1 = f_1, \quad b_1 = g_1$$

$$a_{k+1} = f_{k+1} - E_{A_k} f_{k+1}, \quad b_{k+1} = g_{k+1} - E_{A_k} g_{k+1},$$

$$y_k = \sum_{i=1}^{k} a_i, \quad z_k = \sum_{i=1}^{k} b_i, \quad k \in \mathbf{N}.$$

Since $E_{A_k} g_{k+1} = -E_{A_k f_{k+1}}$, we have $x_n = y_n + z_n$, $n \in \mathbf{N}$. That is why,

$$\bar{x}^{\#} \leq \bar{y}^{\#} + \bar{z}^{\#} \leq \bar{y}^{\#} + \sum_{k=1}^{\infty} |b_k|, \tag{8.16}$$

$$P\bar{y} \leq P\bar{x} + P\bar{z} \leq P\bar{x} + \sum_{k=1}^{\infty} |b_k|. \tag{8.17}$$

Since $|f_k| \leq 2e_{k-1}$, $k \in \mathbf{N}$, then $|a_k| \leq 4e_{k-1}$.

As in theorem 6.3 we can prove an inequality similar to (6.14): for every $a > 1$, and for every $b \in (0, a-1)$ we have

$$mes\{t \in (0,1) : \bar{y}^{\#}(t) > a\lambda, \; \max(P\bar{y}(t), 4e(t)) \leq b\lambda\} \leq$$

$$\leq 2b^2(a-b-1)^{-2} mes\{t \in (0,1) : \bar{y}^{\#}(t) > \lambda)$$

for every $\lambda > 0$.

Now we can use Lemma 6.4 with $a = 2$, $\gamma = 2$ and b such that $0 < b < 1$, and

$$\|\sigma_{4b^2(1-b)^{-2}}\|_E < 2^{-1}.$$

Then, we obtain

$$\|\bar{y}^{\#}\|_E \le C(\|P\bar{y}\|_E + 4\|e\|_E),$$

where C depends only on E.

By combining (8.16) and (8.17) with the obvious inequality $e \le P\bar{x}$, we get

$$\|\bar{x}^{\#}\|_E \le C(\|P\bar{x}\|_E + \|\sum_{k=1}^{\infty} | b_k | \|_E). \tag{8.18}$$

So, it remains to estimate $\sum_{k=1}^{\infty} | b_k |$. Note first of all that by definition

$$\sum_{k=1}^{\infty} | b_k | \le \sum_{k=1}^{\infty} | g_k | + \sum_{k=2}^{\infty} E_{A_{k-1}} | g_k | .$$

If $| d_k(t) | > 2e_{k-1}(t)$, $k = 2, 3, \ldots$, then

$$| d_k(t) | + 2e_{k-1}(t) \le 2 | d_k(t) | \le 2e_k(t).$$

It means that $| g_k | \le 2(e_k - e_{k-1})$, $k = 2, 3, \ldots$, and consequently, we have $\sum_{k=1}^{\infty} | g_k | \le 2e$. To estimate $\| \sum_{k=2}^{\infty} E_{A_{k-1}} | g_k | \|_E$, we use Lemmas 8.9 and 8.10. Finally, we get

$$\|\sum_{k=1}^{\infty} | b_k | \|_E \le 6\|e\|_E \le 6\|P\bar{x}\|_E.$$

The last inequality together with (8.18) completes the proof of the right part of (8.5). The left part is proved similarly.

Let us prove now the necessity of the condition $\lim_{\tau\to 0} \|\sigma_\tau\|_E = 0$. If $\lim_{\tau\to 0} \|\sigma_\tau\|_E > 0$, then by Theorem 1.d.3 for every $n \in \mathbf{N}$ there exist mutually disjoint functions $\{x_i\}_1^{2^n}$ having the same distribution function, such that $\{x_i\}_1^{2^n}$ are 2-equivalent to the unit vector basis of $l_\infty^{2^n}$. We may assume without loss of generality that the functions $\{x_i\}_1^{2^n}$ have the same form as in Theorem 6.3. We suppose, for convenience, that n is an odd number. It is clear that the functions

$$z_1 = x_1 + \dots + x_n,$$

$$z_2 = -x_1 - \dots - x_{2^{n-1}} + x_{2^{n-1}+1} + \dots + x_{2^n},$$

$$z_3 = x_1 + \dots + x_{2^{n-2}} - x_{2^{n-2}+1} - \dots - x_{2^{n-1}},$$

$$z_4 = -x_1 - \dots - x_{2^{n-3}} + x_{2^{n-3}+1} - \dots - x_{2^{n-2}},$$

$$\dots$$

$$z_n = x_1 + x_2 - x_3 - x_4,$$

$$z_{n+1} = -x_1 - x_2$$

form up a martingale-difference sequence, and we have

$$\sup_{1\le k\le n+1} \mid \sum_{i=1}^k z_i \mid \le 2 \sum_{j=1}^{2^n} \mid x_j \mid,$$

$$\left(\sum_{i=1}^{n+1} \mid z_i \mid^2\right)^{1/2} \ge \sqrt{n+1} \mid x_1 \mid .$$

Since $\{x_i\}_1^{2^n}$ are 2-equivalent to the unit vector basis of $l_\infty^{2^n}$, then

$$\| \sup_{1\le k\le n+1} \mid \sum_{i=1}^k z_i \mid \|_E \le 4,$$

$$\|(\sum_{i=1}^{n+1} \mid z_i \mid^2)^{1/2}\|_E \ge \tfrac{\sqrt{n+1}}{2}.$$

These inequalities contradict (8.5). □

Corollary 8.11. *Let E be an r.i. space, $0 < \alpha_E \le \beta_E < 1$. There exist some constants $C_1(E)$ and $C_2(E)$ such that for every martingale $\bar{x}$ we have*

$$C_1(E)\|P\bar{x}\|_E \le \|\bar{x}\|_E \le C_2(E)\|P\bar{x}\|_E. \tag{8.19}$$

Remark 8.12. *The result similar to theorem 6.1 is valid for the Paley function of the arbitrary martingale:*

$$mes\{t \in (0,1) : P\bar{x}(t) > u\} \le \frac{\sqrt{e}}{u}\|\bar{x}\|_{L_1}, \quad u > 0.$$

The constant $\sqrt{e}$ is the best possible one.

Chapter 9

Reproducibility of the Haar system

Let $\{x_n\}_1^\infty$ be a normalized basis of a Banach space X, x_n tends weakly to 0, and X is isomorphic to a subspace of some space Y, with a basis $\{y_k\}_1^\infty$. It is known (see Theorem 1.b.3) that there exists a subsequence $\{x_{n_j}\}_{j=1}^\infty$ of $\{x_n\}_1^\infty$ which is equivalent to a block basis of $\{y_k\}_1^\infty$. It is natural to say in such situations that the subsequence $\{x_{n_j}\}_{j=1}^\infty$ is reproduced as a block basis of $\{y_k\}_1^\infty$. Of particular interest is the case when the above mentioned assertion is valid for the basis $\{x_n\}_1^\infty$ itself. To investigate such situations the following definition is introduced in [161].

Definition 9.1. *A basis $\{x_n\}_1^\infty$ of a Banach space is said to be K-reproducible, for some $K \geq 1$, if, for every isometric embedding of X into a space Y with a basis $\{y_k\}_1^\infty$ and every $\varepsilon > 0$, there exists a block basis $\{z_n\}_1^\infty$ of $\{y_k\}_1^\infty$ which is $(K+\varepsilon)$-equivalent to $\{x_n\}_1^\infty$. When $K = 1$ the basis is said to be precisely reproducible.*

Theorem 9.2. *The Haar basis of every separable r.i. space E is precisely reproducible.*

To prove the theorem we need some preliminary results.

Lemma 9.3. *Let E be an r.i. space, and let $\{\psi_m\}_1^\infty$ be a sequence of functions taking only three values: -1,0,1. Define*

$$A_1 = A_0^0 = A_2 = A_0^1 = \{t \in (0,1) : \psi_1(t) = 1\},$$

and

$$A_m = A_n^k = \{t \in (0,1) : \psi_{2^{n-1}+[\frac{k+1}{2}]}(t) = (-1)^{k+1}\}$$

for $m = 2^n + k$, $(n,k) \in \Omega_1$. Assume that the following conditions are true:

$$1)\ mes(A_n^k) = 2^{-n} \quad for \quad (n,k) \in \Omega_1;$$

$$2)\ A_n^j \cap A_n^i = 0 \quad if \quad i \neq j, \quad n \in \mathbf{N};$$

$$3)\ A_{n+1}^{2k-1} \cup A_{n+1}^{2k} = A_n^k \quad for \quad (n,k) \in \Omega_1.$$

Then a sequence $\{\psi_n\}_1^\infty$ *is isometrically equivalent to the H.s.*

Proof. Let $\{a_{n,k}\}_{(n,k)\in\Omega}$ be a sequence of reals. Then, by properties 1-3, we have, for every $N = 0, 1, \ldots$ and every $\lambda > 0$

$$2^{N+1}\, mes\{t \in (0,1) :\mid \sum_{n=0}^{N} \sum_{k=1}^{2^n} a_{n,k}\psi_{2^n+k}(t) \mid > \lambda\} =$$

$$= card\{k \in \{1, 2, \ldots, 2^{N+1}\} :\mid \sum_{n=0}^{N} \sum_{k=1}^{2^n} a_{n,k}\psi_{2^n+k}(t) \mid > \lambda, t \in A_{N+1}^k\} =$$

$$= card\{k \in \{1, 2, \ldots, 2^{N+1}\} :\mid \sum_{n=0}^{N} \sum_{k=1}^{2^n} a_{n,k}\chi_n^k(t) \mid > \lambda, \quad t \in \Delta_{n+1}^k\} =$$

$$= 2^{N+1}\, mes\{t \in (0,1) :\mid \sum_{n=0}^{N} \sum_{k=1}^{2^n} a_{n,k}\chi_n^k(t) \mid > \lambda\}.$$

So, the functions $\sum_{n=0}^{N} \sum_{k=1}^{2^n} a_{n,k}\psi_{2^n+k}$ and $\sum_{n=0}^{N} \sum_{k=1}^{2^n} a_{n,k}\chi_n^k$ have the same distribution function for every N and every $\{a_{n,k}\}_{(n,k)\in\Omega}$. By the definition of r.i. spaces this means that the system $\{\psi_n\}_1^\infty$ is isometrically equivalent to the H.s. □

Lemma 9.4. *Let E be a separable r.i. space. Then the Rademacher functions tend weakly to 0 in E.*

Proof. Since $E^* = E' \subset L_1$ (see Section 1.d), it suffices to prove that, for every function $x \in L_1$, we have

$$\lim_{n\to\infty} \int_0^1 x(t) r_n(t)\, dt = 0.$$

By Mercer theorem the Fourier coefficients of integrable functions with respect to uniformly bounded orthonormal systems tend to 0. This proves the statement.

Proof of Theorem 9.2. Let E be a separable r.i. space. Assume that T is an isometric embedding of E into a Banach space Y with a basis $\{y_k\}_{k=1}^\infty$, $\{P_N\}_{N=1}^\infty$ is a sequence of the natural projections associated to $\{y_k\}_{k=1}^\infty$,

$$C := C_Y(\{y_k\}_{k=1}^\infty) := \sup_{N\geq 1} \|P_N\|_Y.$$

The assertion of the theorem will be proved if we construct, for any $\varepsilon \in (0,1)$, a sequence $\{\psi_m\}_{m=1}^\infty$ satisfying the following conditions:

1) $\{\psi_m\}_{m=1}^\infty$ complies with the terms of Lemma 9.3;

2) there exists an increasing sequence of positive integers $\{N_j\}_{j=0}^\infty$ such that we have, for any $m \in N$,

$$\|P_{N_{m-1}}(T\psi_m)\|_Y \leq \frac{\varepsilon}{2C(1+\varepsilon)} 2^{-m-1} \|\psi_m\|_E, \tag{9.1}$$

$$\|(I - P_{N_m})(T\psi_m)\|_Y \le \frac{\varepsilon}{2C(1+\varepsilon)} 2^{-m-1} \|\psi_m\|_E. \tag{9.2}$$

Indeed, (9.1) and (9.2) imply that the block basis of $\{y_k\}_{k=1}^{\infty}$ defined by the equations

$$x_m := (P_{N_m} - P_{N_{m-1}})T\psi_m, \quad m \in \mathbf{N},$$

satisfies the following condition:

$$2C \sum_{m=1}^{\infty} \|T\left(\frac{\psi_m}{\|\psi_m\|_E}\right) - x_m\|_Y \le \frac{\varepsilon}{1+\varepsilon}.$$

By Theorem 1.b.2 and Lemma 9.3 this means that $\{x_m\}_{m=1}^{\infty}$ is $(1+\varepsilon)$-equivalent to the H.s. We shall construct $\{\psi_m\}_{m=1}^{\infty}$ and $\{N_j\}_{j=0}^{\infty}$ by induction. Let $\psi_1 = \chi_1$, $N_0 = 0$, N_1 is chosen in such a way that

$$\|(I - P_{N_1})(T\psi_1)\|_Y \le \frac{\varepsilon}{2C(1+\varepsilon)} 2^{-2} \|\psi_1\|_E.$$

By Lemma 9.4 the Rademacher functions tend weakly to 0 in E. Since the projections P_N are completely continuous, there exists a number k_0 such that

$$\|P_{N_1}(Tr_{k_0})\|_Y \le \frac{\varepsilon}{2C(1+\varepsilon)} 2^{-3} \|r_{k_0}\|_E.$$

Put $\psi_2 = r_{k_0}$ and define $N_2 > N_1$ in such a way that (9.2) is valid for $m = 2$. Fix some $n \in N$ and some $k \in \{1, 2, ..., 2^n\}$. Assume that the functions $\psi_1, ..., \psi_{2^n+k-1}$ and the numbers $\{N_j\}_{j=0}^{2^n+k-1}$ are defined with validity of conditions 1 and 2. For each $l \in \{0, 1, 2, ...\}$ let J_l be the set of such $j \in \{2^l+1, 2^l+2, ..., 2^{l+1}\}$ for which $supp\,\chi_j \subseteq \{t \in (0,1) : \psi_{2^{n-1}+[(k+1)/2]}(t) = (-1)^{k+1}\}$. By Lemma 9.4 the sequence of functions $f_l = \sum_{j \in J_l} \chi_j$ tends weakly to 0 in E if $l \to \infty$. That is why there exists an l_0 such that

$$\|P_{N_{2^n+k-1}}(Tf_{l_0})\|_E \le \frac{\varepsilon}{2C(1+\varepsilon)} 2^{-2^n-k-1} \|f_{l_0}\|_E.$$

Put $\psi_{2^n+k} = f_{l_0}$. The integer $N_{2^n+k} > N_{2^n+k-1}$ is defined by (9.2). The functions $\{\psi_m\}_{m=1}^{\infty}$ and the integer $\{N_j\}_{j=0}^{\infty}$ constructed in such manner satisfy conditions 1 and 2. □

Another proof of Theorem 9.2 can be obtained on the basis of the theorem of Liapounoff on convexity of the set of values of finite non-atomic measures (see [163, p. 158]).

Theorem 9.2 and Corollary 1.b.7 imply

Corollary 9.5. *Let Y be a Banach space which has a basis whose unconditional constant K is finite. Then, for every r.i. space E which is isomorphic to a subspace of Y, the Haar system forms an unconditional basis. Moreover, $UC_E(\{\chi_m\}_1^{\infty}) \le Kd$, where $d = \inf\{d(X, Z) : Z$ is a subspace of $Y\}$.*

Remark 9.6. *Corollary 9.5 means, in particular, that a separable r.i. space E has an unconditional basis if and only if the H.s. is also an unconditional basis in that space, i.e.* $0 < \alpha_E \le \beta < 1$.

Theorem 9.7. *Let $\{x_l\}_{l=1}^{\infty}$ be an unconditional basis in an r.i. space E. Then, $\{x_l\}_{l=1}^{\infty}$ is equivalent to the block basis of the H.s. in E.*

Proof. By Theorem 1.b.8 there exists $K \in (0, \infty)$ such that we have, for any $\{a_k\}_1^{\infty}$,

$$K^{-1}\|(\sum_{l=1}^{\infty} | a_l x_l |^2)^{1/2}\|_E \le \|\sum_{l=1}^{\infty} a_l x_l\|_E \le K\|(\sum_{l=1}^{\infty} | a_l x_l |^2)^{1/2}\|_E.$$

Since E is separable, the set of step functions with respect to dyadic intervals is dense in E. That is why, by Theorem 1.b.2, we may assume that

$$x_l = \sum_{k=1}^{2^{n_l}} b_{lk} \kappa_{\Delta_{n_l}^k}, \quad l \in N,$$

where $\{n_l\}_1^{\infty}$ is an increasing sequence of integers. Note that

$$\left(\sum_{k=1}^{2^{n_l}} b_{lk} \kappa_{\Delta_{n_l}^k}\right)^2 = \left(\sum_{k=1}^{2^{n_l}} b_{lk} \chi_{n_l}^k\right)^2.$$

Using Remark 9.6 and Corollary 8.11 we have, for any $\{a_l\}_1^{\infty}$,

$$\|\sum_{l=1}^{\infty}\sum_{k=1}^{2^{n_l}} b_{lk}\chi_{n_l}^k\|_E \approx \|(\sum_{l=1}^{\infty} a_l^2(\sum_{k=1}^{2^{n_l}} b_{lk}\chi_{n_l}^k)^2)^{1/2}\|_E =$$

$$= \|(\sum_{l=1}^{\infty} a_l^2 x_l^2)^{1/2}\|_E \approx \|\sum_{l=1}^{\infty} a_l x_l\|_E.$$

So, $\{x_l\}_1^{\infty}$ is equivalent to the block basis of the H.s. $\{\sum_{k=1}^{n_l} b_{lk}\chi_{n_l}^k\}_{l=1}^{\infty}$ □.

Corollary 9.8. *Every unconditional basis in an r.i. space is reproducible.*

Remark 9.9. If the space E is q-concave for some finite q, then the assertion of Theorem 9.7 is true for every unconditional basic sequence. This follows from Remark 1.b.9. However, in a q-concave r.i. space E, $q < \infty$, an unconditional basic sequence may not be reproducible. For example, for every $p > 1$ there exists the block basis of the H.s. which is equivalent in L_p to the unit vector basis of $(\sum_{n=1}^{\infty} \oplus l_2^n)_p$. Such block basis can be constructed in the following way. Define, for every $n \in N$ and every $i \in \{0, 1, ..., n-1\}$,

$$J_{ni} = \{k \in \{1, 2, ..., 2^{n+i}\} : \Delta_{n+i}^k \subset (2^{-n}, 2^{-n+1})\};$$

$$x_{ni} = \sum_{k \in J_{ni}} \chi_{n+i}^k,$$

$$y_{ni} = x_{ni} / \|x_{ni}\|_{L_p}.$$

It is clear that, for any $n \in \mathbf{N}$, $\{y_{ni}\}_{i=0}^{n-1}$ is equivalent in L_p to $\{r_j\}_{j=1}^{n}$ and, consequently, to the unit vector basis of l_2^n. Here $\{r_j\}_1^{\infty}$ are the Rademacher functions. The whole sequence $\{y_{ni},\ n \in \mathbf{N},\quad i = 0, 1, ..., n-1\}$ is equivalent in L_p to the unit vector basis of $(\sum_{n=1}^{\infty} \oplus l_2^n)_p$. The space $(\sum_{n=1}^{\infty} \oplus l_2^n)_p$ is isomorphic to l_p. This means that, for $p > 1$, $p \neq 2$, the unconditional basic sequence $\{y_{ni},\ n \in \mathbf{N},\ i = 0, 1, ..., n-1\}$ is not reproducible because it is not equivalent to the unit vector basis of l_p, whereas every block basis of the unit vector basis of l_p is equivalent to the unit vector basis of l_p.

Chapter 10

Generalized Haar Systems and Monotone Bases

The purpose of this chapter is to describe monotone bases in r.i. spaces. If any contractive projection P satisfying the condition $P\kappa_{(0,1)} = \kappa_{(0,1)}$ is a conditional expectation, then such description can be given in terms of generalized Haar systems. We start in section 10.a with the characterization of r.i. spaces with the above mentioned property.

10.a. *D*-convexity and *D*-concavity of r.i. spaces

Let E be a separable r.i. space, $S(E) := \{x \in E : \|x\|_E = 1\}$, E' is a dual space, $\langle y, x\rangle := \int_0^1 y(t)x(t)\,dt$.

Definition 10.a.1. *The map $D : E\{0\} \to S(E')$ is called a duality map if for every $x \in E\backslash\{0\}$ $\langle Dx, x\rangle = \|x\|_E$.*

The existence of such map follows from the Hans-Banach theorem.

Example. In L_p, $1 < p < \infty$, the dual map has the following form

$$Dx = sgn\, x \cdot \|x\|_{L_p}^{1-p} \cdot | x |^{p-1} .$$

Definition 10.a.2. *A Banach space E is said to be smooth if the dual map is unique.*

Lemma 10.a.3. *Let E be a smooth separable r.i. space, D is the dual map. If $\lim_{n\to\infty} \|x_n - x\|_E = 0$, $x \neq 0$, then Dx_n converges weakly* to Dx.*

Proof. Assume the contrary. Let x_n converge strongly to x in E, but Dx_n does not converge weakly* to Dx. Passing on, if necessary, to a subsequence we may choose such weak* neighbourhood U of Dx that, for every n, $Dx_n \notin U$. By Alaoglu theorem [131, p.118] a sequence $\{Dx_n\}_1^\infty$ has a weak* limit point f and $\|f\|_{E'} \leq 1$. Besides, we have

$$| \langle f, x\rangle - \|x\|_E | \leq | \langle f, x\rangle - \|x_n\|_E | + \|x_n - x\|_E \leq$$

$$\leq \mid \langle f,x\rangle - \langle Dx_n,x\rangle \mid + \mid \langle Dx_n,x\rangle - \langle Dx_n,x_n\rangle \mid + \|x_n - x\|_E \leq$$
$$\leq \mid \langle f,x\rangle - \langle Dx_n,x\rangle \mid +2\|x_n - x\|_E.$$

That is why $\langle f,x\rangle = \|x\|_E$ and $\|f\|_{E'} = 1$. Since E is smooth, it follows that $f = Dx$. This contradicts the fact that, for every n, $Dx_n \notin U$. □.

Lemma 10.a.4. *Let E be a smooth separable r.i. space. Then*

$$Dx = sgn\, x D|x|, \tag{10.a.1}$$

$$sgn\, Dx = sgn\, x. \tag{10.a.2}$$

Proof. It is clear that

$$\langle Dx,x\rangle = \int_0^1 Dx(t)x(t)\,dt = \|x\|_E = \|\mid x \mid\|_E =$$

$$= \int_0^1 D|x|(t) \mid x(t) \mid dt = \int_0^1 D|x|(t) sgn\, x(t) x(t)\,dt.$$

Furthermore, $\|D|x| sgn\, x\|_{E'} \leq \|D|x|\|_{E'}$. Since E is a smooth space, (10.a.2) follows from (10.a.1). □

Lemma 10.a.5. *Let E be a smooth separable r.i. space. If x is an $\mathcal{A}$-measurable function for some σ-subalgebra $\mathcal{A} \subset \Sigma$, then Dx is also $\mathcal{A}$-measurable.*

Proof. Let x be $\mathcal{A}$-measurable. Hence,

$$\|x\|_E = \int_0^1 x(t)Dx(t)\,dt = \int_0^1 E_{\mathcal{A}}(x(t)Dx(t))\,dt =$$

$$= \int_0^1 x(t)E_{\mathcal{A}}(Dx(t))\,dt.$$

Furthermore, by 1.e.6 we have that $\|E_{\mathcal{A}}(Dx)\|_{E'} \leq \|Dx\|_{E'} = 1$. That is why

$$\|E_{\mathcal{A}}(Dx)\|_{E'} = 1, \quad E_{\mathcal{A}}(Dx) = Dx. \quad \square$$

Lemma 10.a.6. *Let $\{f_n\}_1^\infty$ be a sequence of non-decreasing functions defined on the real line which may have values equal to $+\infty$. Then every subsequence of $\{f_n\}_1^\infty$ has a subsequence converging everywhere, except, maybe, a countable set.*

Proof. We denote for simplicity a subsequence of $\{f_n\}_1^\infty$ as $\{f_n\}_1^\infty$. There exists a sequence $\{n_i\}_1^\infty$ such that $\{f_{n_i}(r)\}_{i=1}^\infty$ converges for every rational r. Let $g(r)$ be the limit of this sequence. The function $g(r)$ is non-decreasing. Extend it continuously from the right on R. Then there exists a countable set $R_1 \subseteq R$ such that g is continuous outside of R_1. It is clear that $f_{n_i}(t) \to g(t)$ if $i \to \infty$ for every $t \notin R_1$. □

Lemma 10.a.7. *Let E be a smooth separable r.i. space. If x and y are equimeasurable, then Dx and Dy are equimeasurable too. Moreover, there exists, for every x, a non-decreasing function f such that $Dx = f \circ x$. The function f depends only on the distribution function of x.*

Proof. By 1.c.1 we have that

$$\|x\|_E = \int_0^1 x(t)Dx(t), dt \le \int_0^1 x^*(t)(Dx)^*(t)\,dt \le \|x^*\|_E\|(Dx)^*\|_{E'} = \|x\|_E.$$

This means that $(Dx)^* = Dx^*$. The last equality together with Lemma 10.a.5 implies that $Dx^* = f \circ x^*$ where f is some non-decreasing function. □

Lemma 10.a.8. *Let E be a smooth separable r.i. space. Then, we have, for any $x, y \in E$, that*

$$\lim_{\varepsilon \to 0}(\|x+\varepsilon y\|_E - \|x\|_E)/\varepsilon = \langle Dx, y\rangle. \tag{10.a.3}$$

Proof. Let $\lambda > 0$. Then

$$\langle Dx, y\rangle = \tfrac{\langle Dx,\lambda y\rangle}{\lambda} = \tfrac{\langle Dx,x\rangle - \|x\|_E + \langle Dx,\lambda y\,\rangle}{\lambda} =$$

$$= \tfrac{Dx,x+\lambda y\rangle - \|x\|_E}{\lambda} \le \tfrac{\|x+\lambda y\|_E - \|x\|_E}{\lambda} \le$$

$$\le \tfrac{\|x+\lambda y\|_E - |\langle D(x+\lambda y),x\rangle|}{\lambda} =$$

$$= \tfrac{\langle D(x+\lambda y),x+\lambda y\rangle - |\langle D(x+\lambda y),x\rangle|}{\lambda} =$$

$$= \tfrac{\langle D(x+\lambda y),x\rangle - |\langle D(x+\lambda y),x\rangle| + \langle D(x+\lambda y),\lambda y\rangle}{\lambda} \le$$

$$\le \langle D(x+\lambda y), y\rangle.$$

By Lemma 10.a.3, we get (10.a.3). □

Definition 10.a.9. *Let E be a smooth separable r.i. space. The norm of E is said to be twice differentiable in 1 if there exists a linear continuous operator $D_1' : E \to E'$ such that*

$$\lim_{\varepsilon \to 0} \|\frac{D(1+\varepsilon x) - 1}{\varepsilon} - D_1'x\|_{E'} = 0 \tag{10.a.4}$$

for every $x \in E$.

Lemma 10.a.10. *Let E be a smooth separable r.i. space with a norm twice differentiable in 1. Then the transformation $(x, y) \to \langle D_1'x, y\rangle$ is bilinear, bounded, symmetric and positive.*

Proof. The first two properties are evident. Let us prove the symmetricity of the transform. Let $\varepsilon > 0$. By (10.a.3) and (10.a.4) there exists a $\lambda_0 > 0$ such that the following inequalities are valid for any $\lambda \in (0, \lambda_0)$:

$$\langle D_1' x, y\rangle \leq \frac{1}{\lambda}\left(\langle D(1+\lambda x), y\rangle - \int_0^1 y(t)\,dt\right) + \varepsilon, \tag{10.a.5}$$

$$\|1+\lambda x\|_E \geq 1 + \lambda \int_0^1 x(t)\,dt - \varepsilon\lambda, \tag{10.a.6}$$

$$\|1+\lambda y\|_E \leq 1 + \lambda \int_0^1 y(t)\,dt - \varepsilon\lambda. \tag{10.a.7}$$

Using the inequalities from the proof of Lemma 10.a.8 we have, for $\lambda \in (0, \lambda_0)$, that

$$< D_1' x, y > \leq \tfrac{1}{\lambda}(< D(1+\lambda x), y > - \textstyle\int_0^1 y(t)\,dt) + \varepsilon \leq$$

$$\leq \tfrac{1}{\lambda}\left(\tfrac{\|1+\lambda x+\lambda y\|_E - \|1+\lambda x\|_E}{\lambda} - \textstyle\int_0^1 y(t)\,dt\right) + \varepsilon \leq$$

$$\leq \tfrac{1}{\lambda}\left(\tfrac{\|1+\lambda x+\lambda y\|_E - 1 - \lambda\int_0^1 x(t)\,dt - \varepsilon\lambda - \lambda\int_0^1 y(t)\,dt}{\lambda}\right) + \varepsilon \leq$$

$$\leq \tfrac{1}{\lambda^2}(\|1+\lambda x+\lambda y\|_E - \|1+\lambda y\|_E + \varepsilon\lambda - \lambda\textstyle\int_0^1 x(t)\,dt - \varepsilon\lambda) + \varepsilon =$$

$$= \tfrac{1}{\lambda}\left(\tfrac{\|1+\lambda x+\lambda y\|_E - \|1+\lambda y\|_E}{\lambda} - \textstyle\int_0^1 x(t)\,dt\right) + \varepsilon \leq$$

$$\leq \tfrac{1}{\lambda}\left(\langle D(1+\lambda(x+y)), x\rangle - \textstyle\int_0^1 x(t)\,dt\right) + \varepsilon.$$

By tending λ to 0, we get that

$$\langle D_1' x, y\rangle \leq \langle D_1' x, x\rangle + \langle D_1' y, x\rangle + \varepsilon.$$

Since $\varepsilon > 0$ is arbitrary, we have that

$$\langle D_1' x, y\rangle - \langle D_1' y, x\rangle \leq \langle D_1' x, x\rangle. \tag{10.a.8}$$

The left part of (10.a.8) linearly depends on y. That is why the inequality may be valid only in the case when

$$\langle D_1' x, y\rangle = \langle D_1' y, x\rangle. \tag{10.a.9}$$

So, the symmetricity of the transform $(x, y) \to \langle D_1' x, y\rangle$ is proved. Moreover, (10.a.8) and (10.a.9) imply positivity of the transform. □

Lemma 10.a.11. *Let E be a smooth separable r.i. space with a norm twice differentiable in 1. Then $E \subset E'$, and there exists a constant $k \geq 0$ such that*

$$\lim_{\varepsilon \to 0} \left\| \frac{1}{\varepsilon}(D(1+\varepsilon x) - \kappa_{(0,1)}) - k(x - \int_0^1 x(t)\,dt) \right\|_{E'} = 0$$

for every $x \in E$. In particular,

$$\|1+\varepsilon x\|_E = 1 + \varepsilon \int_0^1 x(t)\,dt + \frac{\varepsilon^2}{2} k\, var\, x + O(\varepsilon^2), \tag{10.a.10}$$

where $var\, x := \int_0^1 (x(t) - \int_0^1 x(s)\,ds)^2\,dt$.

Proof. By Lemma 10.a.5 the function $D(1+\varepsilon x)$, for every ε, is measurable with respect to the σ-algebra generated by the function x. That is why $D_1'x$ is also x-measurable. Note that $D_1'(\kappa_{(0,1)}) = 0$, since $D(1+\varepsilon\kappa_{(0,1)}) = D(\kappa_{(0,1)}) = \kappa_{(0,1)}$ for every $\varepsilon \neq -1$. By Lemma 10.a.10 we get that

$$\int_0^1 D_1'x(t)\,dt = \int_0^{\prime} D_1'(\kappa_{(0,1)})(t)x(t)\,dt = 0.$$

Thus $D_1'x$ is the x-measurable function with the zero mean.

Fix an arbitrary set $A \in \Sigma$. Then $D_1'\kappa_A = a\kappa_A + b\kappa_{\bar{A}}$, and also

$$a\, mes\,(A) + b(1 - mes\,(A)) = (a-b) mes\,(A) + b = 0.$$

That is why $D_1'\kappa_A = k_A(\kappa_A - mes\,(A))$, where $k_A = a - b$. Let us prove that the constant k_A does not depent on A. Let $D_1'\kappa_B = k_B(\kappa_B - mes\,(B))$. By Lemma 10.a.10 $\langle D_1'\kappa_B, \kappa_A\rangle = \langle \kappa_B, D_1'\kappa_A\rangle$. So, $k_B(mes\,(A \cap B) - mes\,(A) \cdot mes\,(B)) = k_A(mes\,(A \cap B) - mes\,(A) \cdot mes\,(B))$. Thus, $k_A = k_B$ for any dependent set A and set B, for example, for any dyadic intervals not equal to (0,1). This means that k_A does not depend on $A \in \Sigma$. Since the set of step functions is dense in E, we have, for every $x \in E$, that $D_1'x = k(x - \int_0^1 x(t)\,dt)$. In this way we get, for every $x \in E$, that

$$\lim_{\varepsilon \to 0} \| \frac{1}{\varepsilon}(D(1+\varepsilon x) - \kappa_{(0,1)}) - k(x - \int_0^1 x(t)\,dt) \|_{E'} = 0.$$

In particular, it follows from the last equality that $E \subset E'$. By Lemma 10.a.10 we have that $\langle D_1'x, x\rangle \geq 0$. That is why $k \geq 0$.

To prove (10.a.10) we apply the second order Teylor formular to the function $f(\varepsilon) = \|1+\varepsilon x\|_E$. By Lemma 10.a.8 we have that

$$f'(\varepsilon) = \langle D(1+\varepsilon x), x\rangle,$$

$$f''(0) = \langle D_1'x, x\rangle = k\left(\int_0^1 x^2(t)\,dt - (\int_0^1 x(t)\,dt)^2\right) = k\ var\, x. \quad \square$$

Let us consider now the smooth separable r.i. spaces E for which $E' \subset E$. Then we may define the iterations of the duality map. Put

$$D_1 x := Dx, \quad D_{n+1}x := D(D_n x), \quad n \in \mathbf{N}.$$

Lemma 10.a.12. *Let E be a smooth separable r.i. space and $E' \subset E$. Then, a sequence $\{\|D_n x\|_E\}_1^\infty$, for every $x \in E$, is non-decreasing and bounded.*

Proof. Since $\|D_n x\|_{E'} = 1$ for every $n \in N$, the boundedness of the sequence $\{\|D_n x\|_E\}_1^\infty$ follows from the embedding $E' \subset E$. Let us prove that $\|D_n x\|_E \leq \|D_{n+1}x\|_E$ for every $n \in \mathbf{N}$. Indeed,

$$\|D_n x\|_E = \langle D(D_n x), D_n x\rangle \leq \|D_{n+1}x\|_E \|D_n x\|_{E'} = \|D_{n+1}x\|_E. \quad \square$$

Definition 10.a.13. *In an r.i. space E a set F is called E-equiintegrable if*

$$\lim_{c\to\infty} \sup_{x\in F} \|x\kappa_{\{t:|x(t)|>c\}}\|_E = 0.$$

Example. Let $E = L_p, \quad 1 \leq p < 2$. It is clear that the unit ball of $E' = L_q$, where $q = p/(p-1)$, is L_p-equiintegrable.

Theorem 10.a.14. *Let E be a smooth separable r.i. space, $E' \subset E$, and the unit ball of E' is E-equiintegrable.*

Then every subsequence of $\{D_n x\}_1^\infty$, for every $x \in E$, contains a subsequence converging in E. If y is a limit point of $\{D_n x\}_1^\infty$, then $D_2 y = y$.

Proof. If z_n converges almost everywhere to z, and the set $\{z_n\}_1^\infty$ is E-equiintegrable, then $\lim_{n\to\infty} \|z_n - z\|_E = 0$. This assertion is not difficult to prove using Egorov's theorem [131, p.59]. By combining Lemma 10.a.7, Lemma 10.a.6 and the above mentioned fact, we get the proof of the first assertion of theorem 10.a.14.

Let y be a limit point of $\{D_n x\}_1^\infty$. For simplicity of notations we assume that $D_n x$ converges to y in E and almost everywhere. By lemmas 10.a.3 and 10.a.6 there exists a subsequence $\{D_{n_k}\}_{k=1}^\infty$ such that $D_{n_k}x = D_1(D_{n_k-1}x)$ converges to $D_1 y$ almost everywhere. That is why $\|D_1 y\|_E = \lim_{n\to\infty} \|D_{n_k}x\|_E = \|y\|_E$. So, we have that

$$\langle y, D_1 y\rangle = \|y\|_E = \|D_1 y\|_E.$$

Furthermore, $\|y\|_{E'} \leq \inf \lim_{n\to\infty} \|D_n x\|_{E'} = 1$. Thus, $\|y\|_{E'} = 1$ and $D_2 y = y$. $\square$

Remark 10.a.15. Let $E = L_p, \quad 1 < p < \infty$. It is clear that $Dx = sgn\, x \cdot$ $\cdot \mid x \mid^{p-1} / \|x\|_{L_p}^{p-1}$. That is why, if $p \neq 2$,

$$D_2 x = x \Longleftrightarrow Dx = x \Longleftrightarrow x = sgn\, x / \|sgn\, x\|_{L_p'}.$$

Definition 10.a.16. *A smooth separable r.i. space E is called D-concave if*

1) $D_2x = x \Longleftrightarrow x = sgnx/\|sgnx\|_{E'}$;

2) $E' \subset E$, and the unit ball of E' is E-equiintegrable.

If E' is D-concave, then E is called D-convex.

Theorem 10.a.17. *Let E be a D-concave r.i. space. Then a sequence of iterations of the duality map $\{D_nx\}_1^\infty$, for every $x \in E\ \{0\}$, converges in E and almost everywhere to $sgn\, x/\|sgn\, x\|_{E'}$.*

Proof. The assertion follows from Theorem 10.a.14, Lemma 10.a.6 and definition 10.a.16. It is also necessary to keep in mind that, due to Lemma 10.a.4, $sgn\, D_nx = sgn\, x$. □

We are prepared now to answer the following question. For what kind of r.i. spaces is any contractive projection preserving constants a conditional expectation?

Lemma 10.a.18. *Let T be a linear bounded operator in a separable r.i. space E. Denote by $I = \{x \in E : Tx = x\}$. Let $\kappa_{(0,1)} \in I$. If $x \in I$ implies $sgn\, x \in I$, then I is a sublattice of E.*

Furthermore, $I = E(\mathcal{A})$, where $E(\mathcal{A})$ is a subspace of $\mathcal{A}$-measurable functions of E, $\mathcal{A}$ is a σ-algebra of measurable sets A such that $\kappa_A \in I$.

Proof. Let $x \in I$. There exists a sequence $a_n \searrow 0$ as $n \to \infty$, $a_n > 0$, such that, for any $n = 1, 2, \ldots,\quad mes\,\{t : x(t) = a_n\} = 0$. Put $y_n = sgn\,(x - a_n)$. Then $1 + y_n = 2\kappa_{\{t:x(t)>a_n\}} \in I$. It means that $\kappa_{\{t:x(t)>0\}} \in I$. The obtained fact implies that I is a sublattice. Thus, $\mathcal{A} = \{A \in \Sigma : \kappa_A \in I\}$ is a σ-algebra, and obviously $I = E(\mathcal{A})$. □

Theorem 10.a.19. *Let E be a D-concave r.i. space with a norm twice differentiable in 1. Then any contractive projection P satisfying the condition $P\kappa_{(0,1)} = \kappa_{(0,1)}$ is a conditional expectation.*

Proof. Let $I = \{x \in E : Px = x\}$, $\quad P^*$ is a conjugate operator, $I' = \{y \in E' : P^*y = y\}$. Let $x \in I$. Then

$$\langle x, Dx\rangle = \langle Px, Dx\rangle = \langle x, P^*Dx\rangle.$$

Furthermore, $\|P^*\|_{E'} = \|P\|_E = 1$ and that is why $P^*Dx = Dx$. So, $x \in I$ implies $Dx \in I'$. In the same manner one can prove that $y \in I'$ implies $D'y \in I$, where D' is the duality map of E'. Let $y \in I'$. Then, for every ε, $\varepsilon^{-1}(D'(1+\varepsilon y)1) \in I$ and, by Lemma 10.a.11, there exists a constant $k \geq 0$ such that $\varepsilon^{-1}(D'(1+\varepsilon y)-1)$ converges in E to $k(y - \int_0^1 y(t)\,dt)$ if $\varepsilon \to 0$. Thus, $I' \subset I$. In particular, $x \in I$ implies $Dx \in I$. Using Theorem 10.a.17, we get that $sgn\, x \in I$. By Lemma 10.a.18, $I = E(\mathcal{A})$ where $\mathcal{A} = \{A \in \Sigma : \kappa_A \in I\}$. We have, for any $B \in \mathcal{A}$, that

$$\int_0^1 Px(t)\kappa_B(t)\, dt = \int_0^1 x(t)P^*\kappa_B(t)\, dt = \int_0^1 x(t)\kappa_B(t)\, dt.$$

It means that $Px = E_{\mathcal{A}}x$. □

Corollary 10.a.20. *If E is a D-convex r.i. space with a norm twice differentiable in 1, then any contractive projection P satisfying the condition $P\kappa_{(0,1)} = \kappa_{0,1)}$ is a conditional expectation.*

In the case of Orlicz spaces the described method allows to prove the following result.

Theorem 10.a.21. *Let U be an Orlicz function, u is the density function for U, i.e. $U(t) = \int_0^{|t|} u(s)\, ds$. If u is differentiable on R_+ and the derivative $u'(s)$ is strictly increasing to ∞ or strictly decreasing to 0, as $s \to \infty$, then any contractive projector P in L_U satisfying the condition $P\kappa_{(0,1)} = \kappa_{(0,1)}$ is a conditional expection.*

Theorem 10.a.19, Corollary 10.a.20 and Theorem 10.a.21 are generalizations of Ando theorem for $L_p(\nu)$ spaces, $1 < p < \infty$, $p \neq 2$.

Theorem 10.a.22. *Let $1 < p < \infty$, $p \neq 2$. If P is a contractive projection in $L_p(\nu)$, then there exist a measure μ, an isometry $T : L_p(\nu) \to L_p(\mu)$ and a σ-algebra G such that*

$$TPT^{-1} = E_{G,\mu}.$$

Proof. Any closed subspace X of $L_p(\mu)$, where $1 < p < \infty$, contains a function z with the largest support, i.e. for any function $x \in X : supp\, x \subset supp\, z$ [131, p.68].

Let $Im\, P$ be the image of the operator P. If z_0 is an element of $Im\, P$ with the largest support (we denote this by $supp\, P = supp\, z_0$), then we put $z := z_0 + \kappa_{[0,1]\setminus supp\, z_0}$. Define μ as $d\mu =| z |^p\, d\nu$ and let an isometry operator $T : L_p(\nu) \to L_p(\mu)$ be determined by $Tx = x/z$, $x \in L_p(\nu)$. Then $Q = TPT^{-1}$ is a contractive projection in $L_p(\mu)$ and $\kappa_{supp\, Q} = \kappa_{supp\, P} \in Im\, Q$. Thus, by Theorem 10.a.21, $Q|_{L_p(supp\, P)} = E_{G,\mu}$, where G is some σ-algebra. Furthermore, if $p \in (1, \infty)$, then $L_p(\mu)$ is a smooth space and that is why a contractive projection is uniquely determined by its image [80]. So, $Q|_{L_p([0,1]\setminus supp\, P)} = 0$. □

Corollary 10.a.23. *In the conditions of Theorem 10.a.22, if x_1 and x_2 belong to ImP, then $x_1\kappa_{supp\, x_2} \in ImP$.*

Proof. Since $P = T^{-1}E_{G,\mu}T$, we have that

$$x_i = zE_{G,\mu}(\frac{x_i}{z}), \quad i = 1,2. \qquad (10.a.11)$$

That is why

$$zE_{G,\mu}\left(\frac{x_1\kappa_{supp\, x_2}}{z}\right) = z\frac{x_1}{z}E_{G,\mu}(\kappa_{supp\, x_2}) = x_1E_{G,\mu}(\kappa_{supp\, x_2}).$$

Since, by (10.a.11), $\frac{x_2}{z}$ is G-measurable and $supp\; x_2 = \{t : x_2(t)/z(t) \neq 0\}$, we get that $supp\; x_2 \in G$. We obtain finally that

$$zE_{G,\mu}\left(\frac{x_1\kappa_{supp\, x_2}}{z}\right) = x_1\kappa_{supp\, x_2} \in ImP. \quad \square$$

The space $L_1(\nu)$ is not smooth, but any contractive projection in $L_1(\nu)$ preserving constants is a conditional expectation [82]. Furthermore, we have the following analogue of Theorem 10.a.22.

Theorem 10.a.24. *If P is a contractive projection in $L_1(\nu)$, then there exist a measure μ, an isometry $T : L_1(\nu) \to L_1(\mu)$, a σ-algebra G and a nilpotent of norm 1 operator $N : L_1([0,1]\backslash supp\, G, \mu) \to L_1(G,\mu)$ such that*

$$TPT^{-1}(x) = E_{G,\mu}(x) + N(x(1 - \kappa_{supp\, G})).$$

The scheme of the proof of Theorem 10.a.24 is the same as of Theorem 10.a.22, only the operator $Q|_{L_1([0,1]\backslash supp\, P)}$ can be an arbitrary nilpotent contraction because $L_1(\nu)$ is not a smooth space.

10.b. Generalized Haar systems

The Haar system can be naturally generalized to the following notion.

Definition 10.b.1. *Let $(S, \mathcal{F}, \mu)$ be a measure space with a probability non-atomic measure μ. The system of sets $\{A_n^k\}_{(n,k)\in\Omega} \subset \mathcal{F}$ is called a dyadic tree if*

$$A_n^k = A_{n+1}^{2k-1} \cup A_{n+1}^{2k};$$

$$A_{n+1}^{2k-1} \cap A_{n+1}^{2k} = \emptyset$$

for any $(n,k) \in \Omega$.

The generalized Haar system (g.H.s.) $\{H_m\}_1^\infty$ associated with the dyadic tree $\{A_n^k\}_{(n,k)} \in \Omega$ is defined in the following way:

$$H_1(t) = H_0^0(t) = \kappa_{A_0^1}(t), \quad t \in S;$$

$$H_m(t) = H_n^k(t) = \frac{1}{\mu(A_{n+1}^{2k-1})}\kappa_{A_{n+1}^{2k-1}}(t) - \frac{1}{\mu(A_{n+1}^{2k})}\kappa_{A_{n+1}^{2k}}(t), \quad t \in S;$$

where $m = 2^n + k, (n,k) \in \Omega$.

Remark 10.b.2. If $S = [0,1]$, $\mu = mes$ is the Lebesgue measure; $A_n^k = \Delta_n^k$ for $(n,k) \in \Omega$, then the g.H.s. coincides, up to the constant multipliers, with the classical one (the values of the functions in the dyadic rational numbers are of no significance for our consideration).

Remark 10.b.3. It is easy to see that the g.H.s. is a martingale difference sequence and, consequently, it forms a monotone basic sequence in any r.i. space. The closure of a linear span of g.H.s. is a subspace of G-measurable functions, where G denotes σ-algebra, generated by the dyadic tree $\{A_n^k\}_{(n,k)\in\Omega}$. If $S = [0,1], \mathcal{F} = \Sigma = G$ (we remind that Σ is σ-algebra of Lebesgue measurable subsets of [0,1]), $\mu = mes$, then the g.H.s. is a monotone basis in any separable r.i. space. Let us remark also that $G = \Sigma$ if A_n^k are intervals and $\lim_{n\to\infty} \max_{1\le k\le 2^n} mes\,(A_n^k) = 0$. On the other hand, if $G = \Sigma$, then the last equality is valid.

With the help of the g.H.s. it is possible to describe the structure of an arbitrary monotone basis in $L_p(\mu)$, $1 \le p < \infty$, $p \ne 2$.

Theorem 10.b.4. *Let $1 < p < \infty$, $p \ne 2$, $\{x_i\}_1^\infty$ be a normalized monotone basis of $L_p(\nu)$, where ν is a continuous probability measure. Then there exist the measure μ and the isometry $T : L_p(\nu) \to L_p(\mu)$ such that the basis $\{y_i := Tx_i\}_1^\infty$ is the union of non-overlapping subsequences $\{y_m^l\}_{m=1}^\infty, l = 1,2,\ldots$ each of which is a rearrangement of a normalized g.H.s. in $L_p(B_l)$, where $\{B_l\}$ is a system (maybe finite) of disjoint sets covering $S = \cup_l B_l$.*

To prove Theorem 10.b.4 it is necessary to extend Theorem 10.a.22 to a sequence of contractive projections.

Theorem 10.b.5. *Let $\{P_i\}_1^\infty$ be a sequence of contractive projections in $L_p(S,F,\mu)$, $1 < p < \infty$, $p \ne 2$ and also $P_iP_j = P_{\min\{i,j\}}$. Then there exist the measure μ on [0,1], the isometry $T : L_p(\nu) \to L_p(\mu)$ and the sequence of σ-algebras $\mathcal{F}_1 \subset \mathcal{F}_2 \subset \ldots \subset \mathcal{F}$ such that $P_i = T^{-1}E_{\mathcal{F}_i,\mu}T$.*

Proof. Let $z_1 \in Im\,P_1$ and $supp\,z_1 = supp\,P_1$. Let us remind that the equality $supp\,z_1 = supp\,P_1$ means that $supp\,z \subset supp\,z_1$ for every $z_1 \in Im\,P_1$. If $supp\,P_2 \ne supp\,P_1$, then, using Corollary 10.a.23, we can find the function $z_2 \in Im\,P_2$ such that $supp\,z_2 \cap supp\,z_1 = \emptyset$ and $supp\,z_1 \cup supp\,z_2 = supp\,P_2$. If $supp\,P_2 = supp\,P_1$, then we pass to a consideration of P_3. Reasoning in such a way, we obtain the sequence (maybe finite) of disjoint functions and the sequence of natural numbers $n(1) \le n(2) \le \ldots$ with the following properties:

$$supp\,P_i = \cup_{j=1}^{n(i)} supp\,z_j;$$

$$z_j \in Im\,P_i,\ 1 \le j \le n(i);$$

for every $i \in N$.

Define the function $z := \sum_j z_j$, the measure $\mu : d\mu := |z|^p d\nu$, and the isometry $Tx := x/z$. It is clear that the projections $Q_i = TP_iT^{-1}$ satisfy the conditions:

$$Q_i(\kappa_{supp\, Q_i}) = \kappa_{supp\, Q_i};$$

$$Q_iQ_j = Q_{\min\{i,j\}}.$$

By Theorem 10.a.22, there exists the sequence of σ-algebras $\mathcal{F}_1 \subset \mathcal{F}_2 \subset ... \subset \mathcal{F}$ such that $Q_i = E_{\mathcal{F}_i,\mu}$ for every $i \in N$. □

Proof of Theorem 10.b.4. By Theorem 10.b.5, we can suppose that the natural projections $P_i : L_p(\nu) \to [x_1, ..., x_i]$ are equal to $E_{\mathcal{F}_i}$, $i \in N$, where $\mathcal{F}_1 \subset \mathcal{F}_2 \subset ... \subset \mathcal{F}$. Then $Im\, P_i = L_p(\mathcal{F}_i) = [x_1, ..., x_i]$ and, consequently, $\mathcal{F}_i$ is generated by i atoms. Moreover, for every $i \in N$ there are two possibilities: $supp\, P_i = supp\, P_{i-1}$ or $supp\, P_i \neq supp\, P_{i-1}$. In the first case $\mathcal{F}_i$ is obtained from $\mathcal{F}_{i-1}$ by the decomposition into two parts of one of the atoms of $\mathcal{F}_{i-1}$ (denote it by A). It is clear that $supp\, x_i = A$ and $\int_A x_i\, d\nu = 0$. In the second case $\mathcal{F}_i$ is obtained from $\mathcal{F}_{i-1}$ by adding the atom D disjoint with atoms of $\mathcal{F}_{i-1}$. Then $x_i = \pm\kappa_D/(\nu(D))^{1/p}$.

If we denote by x_1^l the functions x_i appearing in the second case, then, including in the sequence $\{x_m^l\}_{m=1}^\infty$ all the functions x_j with $supp\, x_j \subset supp\, x_1^l$, we shall get the representation of the basis $\{x_i\}_{i=1}^\infty$ given in the statement of Theorem 10.b.4. □

Remark 10.b.6. Under the conditions of Theorem 10.b.4 it is possible to obtain as measure μ the Lebesgue measure. The fact is that there exists a mapping ω defined on $\cup_{i=1}^\infty \mathcal{F}_i$ with the following properties:

1) $\omega(A)$ is an interval of [0,1] for every $A \in \cup_{i=1}^\infty \mathcal{F}_i$;
2) if $A \subset B$, then $\omega(A) \subset \omega(B)$;
3) if $A \cap B = \emptyset$, then $\omega(A) \cap \omega(B) = \emptyset$;
4) $\mu(A) = mes\,(\omega(A))$ for every $A \in \cup_{i=1}^\infty \mathcal{F}_i$.

Moreover, we can require that the interval $\omega(\{t : x_i(t) > 0\})$ should be to the left of the interval $\omega(\{t : x_i(t) < 0\})$, $i \in \mathbf{N}$. Such mapping can be continued to the isomorphism of the measure spaces $(S, \mathcal{F}, \mu)$ and $([0,1], \Sigma, m)$.

So, in order to investigate monotone bases in $L_p(\nu)$, $1 < p < \infty$, $p \neq 2$, it is sufficient to study the g.H. systems associated with the dyadic trees of intervals, where the interval A_n^{2k-1} is located to the left of the interval A_n^{2k} for every $n \in N$ and every $k \in \{1, 2, ..., 2^{n-1}\}$.

Monotone bases of L_1 are also constructed on the basis of g.H. systems, but their structure is more complicated due to the larger variety of contractive projections in L_1 (see Theorem 10.a.24).

Theorem 10.b.7. *Let $\{x_i\}_1^\infty$ be a normalized monotone basis of $L_1(\nu)$, and let ν be a continuous probability measure. Then there exist the* *measure μ, the isometry $T : L_1(\nu) \to L_1(\mu)$ and the sequence (maybe finite) of disjoint sets B_l, covering $S = \cup_l B_l$, such that the sequence $\{y_i := Tx_i\}_1^\infty$ is the union of non-overlapping subsequences $\{y_m^l\}_{m=1}^l$ each of which forms together with $\kappa_{B_l}(\mu(B_l))^{-1}$ the g.H.s. normalized in $L_1(B_l)$. Moreover, $y_1^l = C_l\kappa_{B_l} + f_l$, where C_l is a constant, f_l is a linear combination of the elements from $\{y_i\}_1^\infty$ preseding y_1^l and $\|f_l\|_{L_l(\mu)} \leq |C_l|\mu(B_l)$.*

Proof. Let $\{P_i\}_1^\infty$ be the sequence of the natural projections for the basis $\{x_i\}_1^\infty$. Using Theorem 10.a.24 and the method of reasoning from Theorem 10.b.4, we can define the measure μ, the isometry $T : L_1(\nu) \to L_1(\mu)$, and the sequence of σ-algebras $\mathcal{F}_1 \subset \mathcal{F}_2 \subset \ldots \subset \mathcal{F}$ in such a way that the projections $Q_i := TP_iT^{-1}$ have the following form $Q_iz = E_{\mathcal{F}_i}(z) + N_i(z\kappa_{[0,1]\setminus supp\,\mathcal{F}_i})$, where N_i is a nilpotent operator from $L_1([0,1]\setminus supp\,\mathcal{F}_i, \mu)$, into $L_1([0,1], \mathcal{F}_i, \mu)$, $\|N_i\| = 1$. Let us fix $i \in \mathbf{N}$. Two cases are possible: $supp\,Q_i = supp\,Q_{i-1}$ or $supp\,Q_i \neq supp\,Q_{i-1}$. In the first case $\mathcal{F}_i$ is obtained from $\mathcal{F}_{i-1}$ by decomposition of one of the atoms; and $y_i = Tx_i$ is a normalized generalized Haar function with the support equal to this atom. In the second case $\mathcal{F}_i$ is obtained from $\mathcal{F}_{i-1}$ by adding the atom A disjoint with $supp\,\mathcal{F}_{i-1}$. Easy calculations show that

$$Q_{i-1}(\kappa_A - N_{i-1}(\kappa_A)) = 0,$$

$$Q_i(\kappa_A - N_{i-1}(\kappa_A)) = \kappa_A - N_{i-1}(\kappa_A).$$

Consequently, $y_i = C(\kappa_A - N_{i-1}(\kappa_A))$, where C is a constant, $N_{i-1}(\kappa_A)$ is an $\mathcal{F}_{i-1}$-measurable function and $\|N_{i-1}(\kappa_A)\|_{L_1(\mu)} \leq \|\kappa_A\|_{L_1(\mu)} = \mu(A)$. □

In the case of r.i. spaces we have the following

Theorem 10.b.8. *Let E be a D-concave or D-convex r.i. space with a norm twice differentiable in one. Then every normalized monotone basis $\{x_i\}_1^\infty$ with $x_1 = \kappa_{[0,1]}$ is a rearrangement of some g.H.s. normalized in E.*

The proof of Theorem 10.b.8 is similar to that of Theorem 10.b.4: it is necessary only to use Theorem 10.a.19 or Corollary 10.a.20 instead of Theorem 10.a.22.

Let us investigate the equivalence of g.H. systems in L_p. At first we shall consider the case : $1 < p < \infty$, $p \neq 2$. We need the following

Definition 10.b.9. *Let $\{h_n^k\}_{(n,k)\in\Omega}$ be g.H.s. normalized in $L_p, 1 < p < \infty$. The functions*

$$\rho_n = 2^{-\frac{n}{p}} \sum_{k=1}^{2^n} h_n^k, \quad n = 0, 1, \ldots$$

are called the generalized Rademacher functions (g.R.f.'s).

Theorem 10.b.10. *In $L_p(0,1)$, $1 < p < \infty$, $p \neq 2$, there exist non-equivalent to each other g.H. systems.*

Proof. Let us construct the g.H.s. $\{h_n^k\}_{(n,k)\in\Omega}$, normalized in L_p, for which the g.R.f.'s $\{\rho_{2n}\}_{n=0}^{\infty}$ are equivalent to the unit vector basis of l_p. Due to the Khinchin inequality the classical Rademacher functions are equivalent to the unit vector basis of l_2. Consequently, $\{h_n^k\}_{(n,k)\in\Omega}$ will be non-equivalent to the classical H.s.

In order to realize this idea we need some auxiliary facts.

Lemma 10.b.11. *Let $1 \leq p < \infty, \{A_n\}_0^\infty$ is a sequence of disjoint subsets of [0,1]; $\{y_n\}_0^\infty$ is a sequence of functions with the following properties: $\operatorname{supp} y_n \subset A_n, 0 < a \leq \|y_n\|_{L_p} \leq b < \infty, n = 0,1,\ldots$ Then the sequence $\{y_n\}_0^\infty$ is equivalent in L_p to the unit vector basis of l_p.*

Proof. For an arbitrary sequence of scalars $\{c_n\}_0^\infty$ we have

$$a(\textstyle\sum_0^\infty \mid c_n \mid^p)^{1/p} \leq \left(\sum_0^\infty \mid c_n \mid^p \int_0^1 \mid y_n(t) \mid^p dt\right)^{1/p} =$$
$$= \left(\textstyle\int_0^1 \mid \sum_0^\infty c_n y_n(t) \mid^p dt\right)^{1/p} \leq b(\sum_0^\infty \mid c_n \mid^p)^{1/p}.$$

□

Lemma 10.b.12. *Let $1 \leq p < \infty$, $\{y_n\}_0^\infty$ is a sequence of normalized functions. If there exists a sequence of disjoint sets $\{A_n\}_0^\infty$ such that*

$$\int_{A_n} \mid y_n(t) \mid^p dt \geq 1 - 4^{-(n+1)p},$$

then $\{y_n\}_0^\infty$ is equivalent to the unit vector basis of l_p.

Proof. It is easy to see that, for every $n = 0, 1, \ldots$,

$$\|y_n - y_n\kappa_{A_n}\|_{L_p} = \left(1 - \int_{A_n} \mid y_n(t) \mid^p dt\right)^{1/p} \leq 4^{-(n+1)}.$$

By Theorem 1.b.2, $\{y_n\}_0^\infty$ is equivalent to $\{y_n\kappa_{A_n}\}_0^\infty$. The last sequence is equivalent to the unit vector basis of l_p by virtue of Lemma 10.b.11. □

Lemma 10.b.13. *Let $p \in (1,\infty)$, $s \in (0,\infty)$, $x_s(t) = a_s\kappa_{A_s}(t) - b_s\kappa_{B_s}(t)$, $t \in (0,1)$, $a_s > 0$, $b_s > 0$, $A_s \cap B_s = \emptyset$, $\operatorname{mes}(B_s)(\operatorname{mes}(A_s))^{-1} = s$, $\|x_s\|_{L_p} = 1$, $\int_0^1 x_t\, dt = 0$. Then*

$$\lim_{s\to 0} \|x_s\kappa_{B_s}\|_{L_p} = 1.$$

Proof. It is easy to calculate that

$$a_s = (mes(A_s))^{-1}\left[(mes(A_s))^{1-p} + (mes(B_s))^{1-p}\right]^{-1/p},$$

$$b_s = (mes(B_s))^{-1}\left[(mes(A_s))^{1-p} + (mes(B_s))^{1-p}\right]^{-1/p}.$$

That is why

$$\|x_s\kappa_{B_s}\|_{L_p} = (mes(B_s))^{\frac{1-p}{p}}\left[(mes(A_s))^{1-p} + (mes(B_s))^{1-p}\right]^{-1/p} =$$

$$= (s^{p-1}+1)^{-1/p}. \quad \square$$

We are prepared now for the construction of the g.H.s. $\{h_n^k\}_{(n,k)\in\Omega}$ mentioned at the begining of the proof of Theorem 10.b.10.

Let $h_0^0 = 1$, $h_0^1 = \chi_0^1$, $h_1^1 = \chi_1^1$, $h_1^2 = \chi_1^2$, $B_0 = [0,1]$. The construction will be carried out by induction. Let us suppose that, for some $N \in \mathbf{N}$, we define the functions h_n^k, $n = 0,1,2,...,2N-1,$; $1 \le k \le 2^n$; and the sets B_j, $j = 0,1,...,N-1$, with the following property: for every $j = 0,1,...,N-1$,

$$\int_{B_j\setminus\cup_{l=j+1}^{N-1}B_l} |\,\rho_{2j}(t)\,|^p\, dt > 1 - 4^{-(j+1)p},$$

where $\rho_{2j} = 2^{-\frac{2j}{p}}\sum_{k=1}^{2^{2j}} h_{2j}^k$ are g.R.f.'s. Let $\{A_{2N}^k\}_{k=1}^{2^{2N}}$ be the elements of the dyadic tree. For every $k = 1,2,...,2^{2N}$, decompose A_{2N}^k into two intervals A_{2N+1}^{2k-1} and A_{2N+1}^{2k} in such a way that

1) $\|h_{2N}^k\kappa_{A_{2N+1}}^{2k}\|_{L_p}^p > 1 - 4^{-(N+1)p}$;

2) for every $j = 0,1,...,N-1$,

$$\int_{B_j\setminus\cup_{l=j+1}^{N}B_l} |\,\rho_{2j}(t)\,|^p\, dt > 1 - 4^{-(j+1)p},$$

where $B_N := \cup_{k=1}^{2^{2N}} A_{2N+1}^{2k}$.

The first condition is possible to fulfill due to Lemma 10.b.13, the measure of B_N being as small as necessary for the second condition.

The intervals A_{2N+1}^k, $k = 1,...,2^{2N+1}$ are decomposed into two equal intervals A_{2N+2}^{2k-1} and A_{2N+2}^{2k}. This finishes the step of induction. By the construction,

$$\lim_{n\to\infty}\max_{1\le k\le 2^n} mes\,(A_n^k) = 0.$$

Consequently, the g.H.s. $\{h_n^k\}_{(n,k)\in\Omega}$ associated with the dyadic tree $\{A_n^k\}_{(n,k)\in\Omega}$ is a basis of L_p (see Remark 10.b.3). Moreover,

$$\int_{C_N} |\,\rho_{2N}(t)\,|^p\, dt \ge 1 - 4^{-(N+1)p}, \quad N = 0,1,2,...,$$

where $C_N := B_N\setminus\cup_{l=N+1}^{\infty} B_l$. By Lemma 10.b.22, the sequence $\{\rho_{2N}\}_{N=0}^{\infty}$ is equivalent to the unit vector basis of l_p. $\square$

Corollary 10.b.14. *In L_p, $1 < p < \infty$, $p \neq 2$, there exists an uncountable set of pairwise non-equivalent g.H.systems.*

Proof. Fix the sequence of scalars $\{t_n\}_1^\infty$ such that: $t_1 = 0$; $t_{n+1} > t_n$, $n \in \mathbf{N}$; $\lim_{n\to\infty} t_n = 1$. Then, for any sequence of signs $\{\varepsilon_n\}_1^\infty$, $\varepsilon_n = \pm 1$, it is possible to construct the dyadic tree $\{A_n^k\}_{(n,k)\in\Omega}$ and the g.H.s. $\{h_n^k\}_{(n,k)\in\Omega}$, associated with it, in such a way that, for any $n \in \mathbf{N}$, $A_n^{2^n-1} = (t_n, t_{n+1})$, $A_n^{2^n} = (t_{n+1}, 1)$, and the system $\{h_n^k : supp\, h_n^k \subset (t_l, t_{l+1})\}$ is equivalent to the classical H.s. for $\varepsilon_l = 1$; but, for $\varepsilon_l = -1$, the same system is equivalent to the g.H.s. defined in Theorem 10.b.10. It is obvious that for different $\{\varepsilon_n\}_1^\infty$ we will obtain non-equivalent to each other g.H.systems. □

Let us investigate the equivalence of g.H. systems in L_1.

Definition 10.b.15. *Let $\{h_n^k\}_{(n,k)\in\Omega}$ be the g.H.s. The sequence $h_0^0, h_0^1, h_1^{k(1)}$, $h_2^{k(2)}, \ldots, h_n^{k(n)}$ is called a chain if, for every $n \in N$,*

$$supp\, h_{n+1}^{k(n+1)} \subset supp\, h_n^{k(n)}.$$

Lemma 10.b.16. *Two g.H.systems $\{h_n^k\}_{(n,k)\in\Omega}$ and $\{g_n^k\}_{(n,k)\in\Omega}$ are equivalent in L_1 if and only if there exists a constant K such that every chain from $\{h_n^k\}_{(n,k)\in\Omega}$ is K-equivalent to the corresponding chain from $\{g_n^k\}_{(n,k)\in\Omega}$.*

Proof. Let $\{A_n^k\}_{(n,k)\in\Omega}$ be the dyadic tree associated with $\{h_n^k\}_{(n,k)\in\Omega}$. Put $X_N = [h_n^k, (n,k) \in \Omega, n < N]$, $N \in \mathbf{N}$. It is evident that $X_N = [\kappa_{A_N^k}, \ 1 \leq k \leq 2^N]$, for $N \in \mathbf{N}$, and every function $x \in X_N$ with $\|x\|_{L_1} = 1$ is a convex linear combination of the functions $\pm(mes(A_N^k))^{-1}\kappa_{A_N^k}$, $1 \leq k \leq 2^N$. That is why any linear operator on X_N, $N \in \mathbf{N}$, attains its norm on $(mes(A_N^k))^{-1}\kappa_{A_N^k}$, $1 \leq k \leq 2^N$. It remains to remark that the function $(mes(A_N^k))^{-1}\kappa_{A_N}^k$ can be represented as a linear combition of the elements of the chain containing h_N^k. □

For convenience we shall consider only the chain $h_0^0, h_0^1, h_1^1, h_2^1, \ldots, h_n^1, \ldots$, although the results can be applied to any chain.

Theorem 10.b.17. *Let the sequences $\{h_n^k\}_{(n,k)\in\Omega}$ and $\{g_n^k\}_{(n,k)\in\Omega}$ be g.H. systems normalized in L_1 and associated with the dyadic trees $\{A_n^k\}_{(n,k)\in\Omega}$ and $\{B_n^k\}_{(n,k)\in\Omega}$ respectively. Let T be a linear operator defined by the equations $Th_n^k = g_n^k$ for $(n,k) \in \Omega$. Put*

$$\begin{aligned} a_n &= \textstyle\int_{B_n^1} T\left((mes(A_n^1))^{-1}\kappa_{A_n^1}\right)(s)\, ds, \\ b_n &= \textstyle\int_{A_n^1} T\left((mes(B_n^1))^{-1}\kappa_{B_n^1}\right)(s)\, ds, \quad n = 0, 1, 2, \ldots \end{aligned}$$

The chain h_0^0, h_n^1, $n = 0, 1, \ldots$ is equivalent in L_1 to the chain g_0^0, g_n^1, $n = 0, 1, \ldots$ if and only if

$$M := \max(var\, \{a_n\}_0^\infty,\ var\, \{b_n\}_0^\infty) < \infty,$$

where $var\,\{a_n\}_0^\infty := \sum_{n=0}^\infty \mid a_n - a_{n+1} \mid$. *Moreover, the constant of equivalence* K *satisfies inequalities:* $M \le K \le M+3$.

Proof. Let $e_n^k := \left(mes(A_n^k)\right)^{-1} \kappa_{A_n^k}$ for $(n,k) \in \Omega$. Then $e_{n+1}^1 = e_n^1 + 2c_n h_n^1$, $n = 0, 1, ...,$ where $c_n = (mes(A_n^1))^{-1} mes(A_{n+1}^2)$. That is why for every (l,k), $l \le n-1, \quad n \in N$, we have that

$$\int_{B_l^k} Te_n^1(s)\,ds = \int_{B_l^k} Te_{n-1}^1(s)\,ds + 2c_{n-1}\int_{B_l^k} g_{n-1}^1(s)\,ds = \int_{B_l^k} Te_{n-1}^1(s)\,ds.$$

Consequently,

$$\textstyle\int_{B_l^k} Te_n^1(s)\,ds = \int_{B_l^k} Te_e^1(s)\,ds, \quad (l,k) \in \Omega, \quad l \le n-1;$$

$$\textstyle\int_{B_l^2} Te_n^1(s)\,ds = \int_{B_{l-1}^1} Te_n^1(s)\,ds - \int_{B_l^1} Te_n^1(s)\,ds = a_{l-1} - a_l, \quad 0 \le l \le n.$$

The function Te_n^1 is constant on B_n^1 and B_l^2 for $1 \le l \le n$, $n \in \mathbf{N}$. That is why

$$\|Te_n^1\|_{L_1} = \textstyle\sum_{l=1}^n \mid \int_{B_l^2} Te_n^1(s)\,ds \mid + \mid \int_{B_n^1} Te_n^1(s)\,ds \mid =$$

$$= \textstyle\sum_{l=1}^n \mid a_{l-1} - a_l \mid + \mid a_n \mid, \quad n \in \mathbf{N}.$$

Since

$$e_{n+1}^2 = e_n^1 - 2(1-c_n)h_n^1, \quad n = 0, 1, ...,$$

it follows that

$$\|Te_{n+1}^2\|_{L_1} \le \|Te_n^1\|_{L_1} + 2.$$

It remains to remark that any function $x \in [h_0^1, h_n^1,\ n = 0, 1, ...]$, $\|x\|_{L_1} = 1$ can be represented in the following form $x = \sum_{n=1}^\infty \alpha_n e_n^2$, where $\sum_{n=1}^\infty \mid \alpha_n \mid = 1$. □

Theorem 10.b.18. *Let the sequences* $\{h_n^k\}_{(n,k)\in\Omega}$ *and* $\{g_n^k\}_{(n,k)\in\Omega}$ *be the g.H. systems associated with the dyadic trees* $\{A_n^k\}_{(n,k)\in\Omega}$ *and* $\{B_n^k\}_{(n,k)\in\Omega}$ *respectively. Let*

$$c_n := \frac{mes\,(A_{n+1}^2)}{mes\,(A_n^1)}, \quad d_n := \frac{mes\,(B_{n+1}^2)}{mes\,(B_n^1)}, \quad n = 0, 1, ...$$

If

$$M := \max\left(var\,\{\frac{c_n}{d_n}\}_0^\infty, \quad var\,\{\frac{d_n}{c_n}\}_0^\infty\right) < \infty,$$

then the chains $h_0^0, h_n^1, \quad n = 0, 1, ...$ *and* $g_0^1, g_n^1, \quad n = 0, 1, ...$ *are* K*-equivalent in* L_1*, where* $K \le 2M+3$.

Proof. It is easy to calculate that

$$a_n = mes(B_n^1)\{1 + \sum_{l=1}^{n} c_{l-1}/mes(B_l^1)\}, \quad n \in \mathbf{N}; \tag{10.b.1}$$

$$mes(B_n^1)\{1 + \sum_{l=1}^{n} d_{l-1}/mes(B_l^1)\} = 1, \quad n \in \mathbf{N}; \tag{10.b.2}$$

where a_n has the same meaning as in Theorem 10.b.17. That is why the statement of Theorem 10.b.18 follows from the following

Lemma 10.b.19. *Let $\{v_n\}_1^\infty$, $\{w_n\}_1^\infty$ be sequences of scalars, and also $w_n > 0$ for every $n \in \mathbf{N}$. Denote by*

$$V_n := \sum_{k=1}^{n} v_n, \quad W_n := \sum_{k=1}^{n} w_k, \quad n \in \mathbf{N}.$$

Then

$$var\,\{\frac{V_n}{W_n}\}_1^\infty \le var\,\{\frac{v_n}{w_n}\}_1^\infty.$$

Proof. Let $v_k = t_k w_k, \quad k \in \mathbf{N}$. Using the Abel transformation we get that

$$\frac{V_n}{W_n} = \frac{1}{W_n}\sum_{k=1}^{n} t_k w_k = \frac{1}{W_n}\sum_{k=1}^{n-1}(t_k - t_{k+1})W_k + t_n, \quad n \in \mathbf{N}.$$

Consequently,

$$\frac{V_{n+1}}{W_{n+1}} - \frac{V_n}{W_n} = \left(\frac{1}{W_n} - \frac{1}{W_{n+1}}\right)\sum_{k=1}^{n}(t_{k+1} - t_k)W_k, \quad n \in \mathbf{N}.$$

Thus, we have that

$$var\left\{\tfrac{V_n}{W_n}\right\}_1^\infty \le \textstyle\sum_{n=1}^\infty \left(\frac{1}{W_n} - \frac{1}{W_{n+1}}\right)\sum_{k=1}^n \mid t_{k+1} - t_k \mid W_k =$$

$$= \textstyle\sum_{k=1}^\infty \mid t_{k+1} - t_k \mid W_k \sum_{n=k}^\infty \left(\frac{1}{W_k} - \frac{1}{W_{k+1}}\right) \le \{t_k\}_1^\infty.$$

□

Corollary 10.b.20. *In L_1 there exists an uncountable set of pairwise non-equivalent g.H. systems.*

Proof. For any $\alpha \in (0,1)$ there exists the dyadic trees $\{A_n^k\}_{(n,k)\in\Omega}$ such that $mes\,(A_n^1))^{-1} mes\,(A_{n+1}^2) = (n+1)^{-\alpha}, \quad n = 0, 1, \ldots$ Equations (10.b.1), (10.b.2) and Theorem 10.b.17 imply that for different α the corresponding g.H. systems are non-equivalent in L_1. □

Theorem 10.b.21. *Let $1 < p < \infty$. There exists a constant K depending only on p such that every chain of g.H. functions h_0^0, $h_n^{k(n)}$, $n = 0, 1, \ldots$, $\|h_n^{k(n)}\|_{L_p} = 1$, is K-equivalent in $L_p(\mu)$ to the unit vector basis of l_p.*

The proof of this theorem is based on the following

Lemma 10.b.22. *Let $1 \le p \le 2$, $\delta > 0$, $\{x_n\}_1^\infty$ is a normalized unconditional basic sequence in L_p. Then, for any sequence of scalars $\{a_n\}_1^\infty$, we have that*

$$\|\sum_{n=1}^\infty a_n x_n\|_{L_p} \le UC_{L_p}(\{x_n\}_1^\infty)(\sum_{n=1}^\infty |a_n|^p)^{1/p}.$$

If there exist disjoint sets $\{B_n\}_1^\infty$ such that $\|x_n \kappa_{B_n}\|_{L_p} \ge \delta$ for $n \in \mathbf{N}$, then

$$\delta\,(UC_{L_p}(\{x_n\}_1^\infty))^{-1}(\sum_{n=1}^\infty |a_n|^p)^{1/p} \le \|\sum_{n=1}^\infty a_n x_n\|_{L_p}.$$

Proof. By Remark 1.b.9 and imbedding $l_p \subset l_2$, we get that

$$\|\sum_{n=1}^\infty a_n x_n\|_{L_p} \le UC_{L_p}(\{x_n\}_1^\infty)\|\left(\sum_{n=1}^\infty |a_n x_n|^2\right)^{1/2}\|_{L_p} \le$$

$$\le UC_{L_p}(\{x_n\}_1^\infty\|\left(\sum_{n=1}^\infty |a_n x_n|^p\right)^{1/p}\|_{L_p} =$$

$$= UC_{L_p}(\{x_n\}_1^\infty)\left(\sum_{n=1}^\infty |a_n|^p\right)^{1/p}.$$

On the other hand,

$$\|\sum_{n=1}^\infty a_n x_n\|_{L_p} \ge \left(UC_{L_p}(\{x_n\}_1^\infty)\right)^{-1}\|\left(\sum_{n=1}^\infty |a_n x_n|^2\right)^{1/2}\|_{L_p} \ge$$

$$\ge \left(UC_{L_p}(\{x_n\}_1^\infty)\right)^{-1}\|\sum_{n=1}^\infty a_n x_n \kappa_{B_n}\|_{L_p} \ge$$

$$\ge \left(UC_{L_p}(\{x_n\}_1^\infty)\right)^{-1}\delta\left(\sum_{n=1}^\infty |a_n|^p\right)^{1/p}.$$

□

Proof of Theorem 10.b.21. If $p = 2$ then the statement is evident. Consider the case $1 < p < 2$. For simplicity of notation we shall give the proof for the chain h_0^0, h_n^1, $n = 0, 1, \ldots$ By Remark 10.b.3 and Corollary 8.4, we have that

$$UC_{L_p}(h_0^0,\ h_n^1,\ n = 0, 1, \ldots) \le \frac{1}{p-1}.$$

Put

$$N_1 := \{n = 0, 1, \ldots : \|h_n^1 \kappa_{A_{n+1}^2}\|_{L_p(\mu)} \ge 2^{-\frac{1}{p}}\},$$

$$N_2 := \{n = 0, 1, \ldots : \|h_n^1 \kappa_{A_{n+1}^1}\|_{L_p(\mu)} > 2^{-\frac{1}{p}}\},$$

where $\{A_n^k\}_{(n,k)\in\Omega}$ is the dyadic tree associated with $\{h_n^k\}_{(n,k)\in\Omega}$, μ is a continuous probability measure. By Lemma 10.b.22, the sequence $\{h_n^1\}_{n\in N_1}$ is $(2^{1/p}/(p-1))$-equivalent in $L_p(\mu)$ to the unit vector basis of l_p. Consider $\{h_n^1\}_{n\in N_2}$. Easy calculations show that $\mu(A_{n+1}^2) > \mu(A_{n+1}^1)$ for $n \in N_2$. Denote by $l(n) := \min\{l : l > n,\ l \in N_2\}$, $B_n := A_{n+1}^1 \backslash A_{l(n)+1}^1$ for $n \in N_2$. Then $\|h_n^1 \kappa_{B_n}\|_{L_p} > 2^{-2/p}$, $n \in N_2$. Consequently, by Lemma 10.b.22 the sequence $\{h_n\}_{n\in N_2}$ is $(2^{2/p}/(p-1))$-equivalent in $L_p(\mu)$ to the unit vector basis of l_p. The case $2 < p < \infty$ is proved by duality argument.

Chapter 11

Haar System Rearrangements

Let $n \geq 0$ and π_n be a rearrangement of $\{0, 1\}$ if $n = 0$ and $\{1, 2, ..., 2^n\}$ if $n \geq 1$. A sequence $\pi = \{\pi_0, \pi_1 ...\}$ generates the operator

$$T_\pi\left(\sum_{(n,k)\in\Omega} c_{n,k}\chi_n^k\right) = \sum c_{n,k}\chi_n^{\pi_n(k)}$$

on the Haar polynomials. By the Parseval equality T_π is isometric in L_2. Clearly, every operator T_π is not bounded from L_p into L_q if $1 \leq p < q \leq \infty$. We shall investigate the boundedness of T_π from L_p into L_q under the assumption $1 \leq q \leq p \leq \infty$. Let $\Omega(0)$ be the set of finite sequences of Ω and G be the cone of functions

$$\sum_{(n,k)\in\Omega(0)} d_{n,k} f_{n,k}(t),$$

where $d_{n,k} \geq 0,\ f_{n,k}(t) = \kappa_{\Delta_n^k(t)}$. Let us consider the operator S_π :

$$S_\pi\left(\sum_{(n,k)\in\Omega(0)} d_{n,k} f_{n,k}\right) = \sum d_{n,k} f_{n,\pi_n(k)},$$

acting in G. It was proved in Chapter 5 that the H.s. is an unconditional basis in L_p $(1 < p < \infty)$ and

$$\frac{1}{A_p}\left\|\left(\sum_{(n,k)\in\Omega} c_{n,k}^2 f_{n,k}\right)^{\frac12}\right\|_{L_p} \leq \left\| \sum_{(n,k)\in\Omega} c_{n,k}\chi_n^k\right\|_{L_p} \leq$$

$$\leq A_p\left\|\left(\sum_{(n,k)\in\Omega} c_{n,k}^2 f_{n,k}\right)^{\frac12}\right\|_{L_p}. \tag{11.1}$$

If $2 \leq q \leq p < \infty$, it follows that

$$\frac{1}{A_pA_q}\|S_\pi\|^{\frac12}_{\frac p2,\frac q2} \leq \|T_\pi\|_{p,q} \leq A_pA_q\|S_\pi\|^{\frac12}_{\frac p2,\frac q2}.$$

Note that the operator S_π norm is defined with respect to the cone G.

Spaces $L_{p,r}$ (Section 1.d) have an analogous property. Let $2 < q \le p < \infty$, $2 \le r < \infty$. We have

$$\|T_\pi\|_{L_{p,r},L_{q,r}} \approx \|S_\pi\|^{\frac{1}{2}}_{L_{p/2,r/2},L_{q/2,r/2}}. \tag{11.2}$$

The operator S_π is positive. It has a more simple structure than T_π.

Denote by $\Pi = \{\pi\}, \pi = \{\pi_0, \pi_1 \ldots\}$. Let $\mathfrak{N}$ be a set of subsets $e \subset [0,1]$ such that e is the finite union of disjoint intervals $\Delta_n^k, (n,k) \in \Omega$. We shall distinguish between Δ_n^k and $\Delta_{n+1}^{2k-1} \cup \Delta_{n+1}^{2k}$.

If $m = 2^n + k,\ (n,k) \in \Omega$, we put

$$\pi(\Delta_m) = \pi(\Delta_n^k) = \cup_{(i,j)\in O(n,k)} \Delta_i^{\pi_i(j)}, \tag{11.3}$$

where

$$O(n,k) = \{(i,j) : (i,j) \in \Omega,\ \Delta_i^j \subset \Delta_n^k\}.$$

Let r be an integer, $\Delta_{n_s} (1 \le s \le r)$ be disjoint intervals and

$$e = \cup_{s=1}^r \Delta_{n_s}.$$

Then we define

$$\pi(e) = \cup_{s=1}^r \pi(\Delta_{n_s}). \tag{11.4}$$

If $\pi^{-1} = (\pi_0^{-1}, \pi_1^{-1} \ldots)$, then $\pi^{-1}(e)$ is defined by (11.3), (11.4).

Theorem 11.1. *Let $2 < q \le p < \infty$, $\pi \in \Pi$ and T_π is bounded from $L_{p,1}$ into $L_{q,\infty}$. Then there exists $C > 0$ such that, for every $A \in \mathfrak{N}$,*

$$mes\,\pi^{-1}(A) \le C(mes\,A)^{1-\theta}, \tag{11.5}$$

where

$$\Theta = \frac{2(p-q)}{q(p-2)}.$$

Proof. Given $A \in \mathfrak{N}$, we choose $B \in \mathfrak{N}$, $B \subset \pi^{-1}(A)$. Let $B = \cup_i \Delta_{m(i)}$, and $\Delta_{m(i)}$ are disjoint intervals. Consider the functions

$$x(t) = \sum_i \chi_{m(i)}, \quad f(t) = \sum_i \chi_{\Delta_{\pi(m(i))}}.$$

Clearly,

$$\|x\|_{L_{p,1}} = (mes\,B)^{\frac{1}{p}}.$$

By (7.a.9), there exists a constant $C_1 > 0$ such that

$$\|T_\pi x\|_{L_{q,\infty}} = \|\textstyle\sum_i \chi_{\pi(m(i))}\|_{L_{q,\infty}} \ge$$
$$\ge C_1\|(\textstyle\sum_i (\chi_{\pi(m(i))}(t))^2{}^{\frac{1}{2}}\|_{L_{q,\infty}} = C_1\|f^{\frac{1}{2}}\|_{L_{q,\infty}}. \tag{11.6}$$

It was noted in Section 1.d that the space $L_{2,q}$ is 2-convex. Therefore

$$\|T_\pi x\|_{L_{q,\infty}} \geq C_2 \|f\|^{\frac{1}{2}}_{L_{\frac{q}{2},\infty}}. \tag{11.7}$$

Applying Theorem 1.e.6, we get

$$\begin{aligned} \|f\|_{L_{\frac{q}{2},\infty}} &\geq \|\tfrac{1}{mesA}\textstyle\int_A f(s)\,ds\kappa_A\|_{L_{\frac{q}{2},\infty}} = \\ &= \textstyle\int_A f(s)\,ds(mesA)^{\frac{2}{q}-1} = mesB(mesA)^{\frac{2}{q}-1}. \end{aligned} \tag{11.8}$$

Combining estimates (11.6-8), we get

$$C_2(mes\,B)^{\frac{1}{2}}(mes\,A)^{\frac{1}{q}-\frac{1}{2}} \leq \|T_\pi x\|_{L_{q,\infty}} \leq \|T_\pi\|(mes\,B)^{\frac{1}{p}}.$$

Hence,

$$(mes\,B)^{\frac{1}{2}-\frac{1}{p}} \leq \frac{\|T_\pi\|}{C_2}(mes\,A)^{\frac{1}{2}-\frac{1}{q}}.$$

It means that

$$mes\,\pi^{-1}(A) \leq \left(\frac{\|T_\pi\|}{C_2}\right)^{\frac{2p}{p-2}}(mes\,A)^{1-\Theta}. \quad \square$$

By duality argument, it is easy to obtain an analogous statement for $1 < q \leq p < 2$.

The next elementary Lemma is presented for the fullness of exposition.

Lemma 11.2. *Let* $0 \leq \Theta < 1$, $1 \geq a_1 \geq a_2 \geq \ldots > 0$, *and for some constant* $C > 0$ *and any* $n = 1, 2, \ldots$ *the following inequality*

$$\sum_{k=n}^{\infty} a_k \leq C a_n^{1-\Theta}$$

holds. Then there exist some constants C_1, C_2 *such that*

$$a_n \leq \begin{cases} C_1 n^{-\frac{1}{\Theta}}, & \Theta > 0, \\ a_1 e^{-C_2 n}, & \Theta = 0, \end{cases}$$

for every $n = 1, 2, \ldots$.

Proof. Let $\Theta > 0$. We have

$$n a_{2n} \leq \sum_{k=n}^{2n} a_k \leq C a_n^{1-\Theta}.$$

Denote $a_n n^{\frac{1}{\Theta}} = b_n$. Then

$$b_{2n} \leq 2^{\frac{1}{\Theta}} C b_n^{1-\Theta}.$$

Hence,

$$b_{2^n} \le 2^{\frac{1}{\Theta}} C b_{2^{n-1}}^{1-\Theta} \le \dots \le (2^{\frac{1}{\Theta}} C)^{\sum_{k=0}^{\infty}(1-\Theta)^k} b_1^{(1-\Theta)^n} \le (2^{\frac{1}{\Theta}} C)^{\frac{1}{\Theta}}$$

and

$$a_{2^n} \le 2^{\frac{1}{\Theta^2}} C^{\frac{1}{\Theta}} 2^{-\frac{n}{\Theta}}.$$

Using standard argument, we get

$$a_n \le 2^{\frac{1}{\Theta^2}+\frac{1}{\Theta}} C^{\frac{1}{\Theta}} n^{-\frac{1}{\Theta}}.$$

If $\Theta = 0$, the proof is analogous ([19],p.22) □

Given $\pi \in \Pi$ and $A \in \mathfrak{N}$, denote

$$\lambda(\tau) = mes\{t : S_{\pi\kappa_A(t)} \ge \tau\}.$$

Lemma 11.3. *Let* $0 \le \Theta < 1$, $\pi \in \Pi$, *condition (11.5) is satisfied and* $A \in \mathfrak{N}$. *Then, for every* $\tau > 0$,

$$\lambda(\tau) \le \begin{cases} C_1 \tau^{-\frac{1}{\Theta}}, & \Theta > 0, \\ e^{-C_2\tau} mes\, A, & \Theta = 0, \end{cases}$$

where C_1 *and* C_2 *are the constants from Lemma 11.2.*

Proof. For every $t \in [0,1]$, $S_\pi\kappa_A(t)$ is an integer. Denote

$$r = \max_{0\le t\le 1} S_\pi \kappa_A(t),$$

$$e_k = \{t : S_\pi\kappa_A(t) \ge k\}, \quad k = 1,2,\dots,r.$$

Then $e_1 \supset e_2 \supset \dots \supset e_r$. There exist disjoint sets $g_k \in \mathfrak{N}$, $mes\, g_k = mes\, e_k$, $g_k \subset \pi^{-1}(e_k)$, $k = 1,2,\dots,r$. Therefore

$$\cup_{j=k}^{r} g_j \subset \pi^{-1}(e_k).$$

Hence,

$$\sum_{j=k}^{r} mes\, e_j = \sum_{j=k}^{r} mes\, g_j \le mes\, \pi^{-1}(e_k), \quad k = 1,2,\dots,r.$$

By Lemma 11.2,

$$\lambda(\tau) = mes\, e_n \le \begin{cases} C_1 n^{-\frac{1}{\Theta}}, & \Theta > 0 \\ e^{-C_2 n} mes\, e_1, & \Theta = 0. \end{cases}$$

Since $\lambda(\tau) = \lambda(n)$ for $\tau \in (n-1, n]$ and $\lambda(1) \le mes A$, then

$$\lambda(\tau) \le \begin{cases} C_1 \tau^{-\frac{1}{\Theta}}, & \Theta > 0; \\ e^{-C_2\tau} mes\, A, & \Theta = 0. \end{cases}$$

□

Lemma 11.4. *Suppose that $\pi \in \Pi$ satisfies condition (11.5). Then S_π acts from L_∞ into $L_{\frac{1}{\Theta},\infty}$ if $\Theta > 0$ and from $L_{p,1}$ and L_p for every $p \in (1,\infty)$ if $\Theta = 0$.*

Proof. Let $A \in \mathfrak{N}$. By Lemma 11.3,

$$(S_\pi \kappa_A)^*(t) \le (\frac{t}{C_1})^{-\Theta}.$$

Hence,

$$\|S_\pi \kappa_A\|_{L_{\frac{1}{\Theta},\infty}} = \sup_{0<\tau\le 1} \frac{\int_0^\tau (S_\pi \kappa_A)^*(t)\,dt}{\tau^{1-\Theta}} \le \sup_{0<\tau\le 1} \frac{\int_0^\tau (\frac{t}{C_1})^{-\Theta}\,dt}{\tau^{1-\Theta}} = \frac{C_1^\Theta}{1-\Theta}.$$

It means that, for every $x \in G$

$$\|S_\pi x\|_{L_{\frac{1}{\Theta},\infty}} \le \frac{C_1^\Theta}{1-\Theta}\|x\|_{L_\infty}.$$

The first part of this Lemma is proved.

Suppose that $\Theta = 0$. Fix $p \in (1,\infty)$. Applying the second part of Lemma 11.3 and formula (1.d.2), we get

$$\|S_\pi \kappa_e\|_{L_p} = (p\int_0^\infty \lambda(\tau)\tau^{p-1}\,d\tau)^{\frac{1}{p}} \le (p\int_0^\infty mes\, A e^{-C_2\tau}\tau^{p-1}\,d\tau)^{\frac{1}{p}} =$$

$$= (p\int_0^\infty e^{-C_2\tau}\tau^{p-1}\,d\tau)^{\frac{1}{p}}(mes A)^{\frac{1}{p}}$$

It was noted in Section 1.d that, if a convex functional is uniformly bounded on the set of characteristic functions in $L_{p,1}$, then it is uniformly bounded on the set of step functions. Therefore, if $x \in G$, then

$$\|S_\pi x\|_{L_p} \le (p\int_0^\infty e^{-C_2\tau}\tau^{p-1}\,d\tau)^{\frac{1}{p}}\|x\|_{L_{p,1}}. \quad \Box$$

Theorem 11.5. *Let $1 < p < \infty$, $p \ne 2$, $\pi \in \Pi$. An operator T_π is bounded in L_p if there exists a constant $C >$ such that, for every $A \in \mathfrak{N}$,*

$$mes\,\pi(A) \le C mes\, A, \quad 1 < p < 2, \tag{11.9}$$

$$mes\,\pi^{-1}(A) \le C mes\, A, \quad 2 < p < \infty. \tag{11.10}$$

Proof. The equality $T_\pi^* = T_{\pi^{-1}}$ shows that the cases $1 < p < 2$ and $2 < p < \infty$ are dual. Consequently, it is sufficient to consider only the case $2 < p < \infty$. We prove that under condition (11.10) the operator S_π is bounded in L_p for every $p \in (1,\infty)$. Let $1 < p < q < \infty$. By Lemma 11.4, the operator

S_π is bounded from $L_{q,1}$ into L_q. It is evident that S_π is isometric in L_1. Recall (see Section 1.e) that

$$K(t,x,L_1,L_{q,1}) = \inf_{x=x_0+x_1}(\|x_0\|_{L_1} + t\|x_1\|_{L_{q,1}}), \quad t>0. \tag{11.11}$$

It is well known [23] that the infimum in (11.11) is attained on a pair (x_0, x_1) such that $x(t) = \min(\lambda, x(t))$ for some $\lambda > 0$. Therefore, if $x \in G$, then $x_0, x_1 \in G$. Hence,

$$K(t, S_\pi x, L_1, L_q) \le \|S_\pi\|_{L_{q,1},L_q} K(t,x,L_1,L_{q,1}).$$

Applying the real interpolation method for the pair $(L_1, L_{q,1})$ and the reitaration theorem (Section 1.e), we get that S_π acts in L_p .

Theorem 11.6. *Let $1<q<p<2$, $1\le r\le 2$ or $2<q<p<\infty$, $2\le r\le\infty$. An operator T_π is bounded from $L_{p,r}$ into $L_{q,r}$ iff there exists a constant $C>0$ such that, for every $A\in\mathfrak{M}$,*

$$mes\,\pi(A) \le C(mes\,A)^{1-\Theta}, \quad 1<q<p<2$$

or

$$mes\,\pi^{-1}(A) \le C(mes\,A)^{1-\Theta}, \quad 2<q<p<\infty, \tag{11.12}$$

where

$$\Theta = \begin{cases} \frac{p-q}{p(2-q)}, & 1<q<p<2, \\ \frac{p-q}{q(p-2)}, & 2<q<p<\infty. \end{cases}$$

Proof. By duality, it is sufficient to consider only the second case. The boundedness of T_π from $L_{p,r}$ into $L_{q,r}$ implies the boundedness T_π from $L_{p,1}$ into $L_{q,\infty}$. Applying Theorem 11.1, we get that condition (11.12) is satisfied.

To prove the second part of this Theorem it is sufficient to repeat the proof of Theorem 11.5. □

Now we present a simple sufficient condition in order that T_π should be bounded in L_p ($p\ne 2$). Note that conditions (11.9) and (11.10) may be written in a more simple form

$$mes\,(\cup_{r=n}^{\infty}\cup_{2^{r-n}(k-1)<s\le 2^{r-n}k}\Delta_r^{\pi_r(s)}) \le C2^{-n}, \tag{11.13}$$

$$mes(\cup_{r=n}^{\infty}\cup_{2^{r-n}(k-1)<s\le 2^{r-n}k}\Delta_r^{\pi_r^{-1}(s)}) \le C2^{-n}. \tag{11.14}$$

Indeed,

$$\pi(\Delta_n^k) = \cup_{r=n}^{\infty}\cup_{2^{r-n}(k-1)<s\le 2^{r-n}k}\Delta_r^{\pi_r(s)}.$$

Therefore (11.9) implies (11.13). Conversely, if (11.13) is valid for every interval Δ_n^k, then summing up these inequalities we obtain (11.9) with same constant.

Corollary 11.7. *Let M be an integer. If, for every $(n,k) \in \Omega$,*

$$| \pi_n(k) - k | \leq M,$$

then $\{\chi_n^k\}$ and $\{\chi_n^{\pi_n(k)}\}$ are equivalent in L_p, $1 < p < \infty$.

Indeed, conditions (11.13) and (11.14) are satisfied where $C = 2M + 1$. Note that the C.Fefferman - E.Stein work [90] leads to Corollary 11.7 too.

Corollary 11.8. *Let M be an integer. If, for every $(n,k) \in \Omega$,*

$$card\{k : \pi_n(k) \neq k\} \leq M,$$

then $\{\chi_n^k\}$ and $\{\chi_n^{\pi_n}(k)\}$ are equivalent in L_p, $1 < p < \infty$.

In this case the inequailities (11.13) and (11.14) are fulfilled, where $C = \log_2 M + 2$.

These corollaries may be united. The sequences $\{\chi_n^k\}$ and $\{\chi_n^{\pi_n(k)}\}$ are equivalent in L_p $(1 < p < \infty)$ if

$$\sup_n \operatorname{card}\{k : |\pi_n(k) - k| \geq M\} \leq M$$

for some integer M. By Theorem 1.e.4, under this assumption, $\{\chi_n^k\}$ and $\{\chi_n^{\pi_n}(k)\}$ are equivalent in any r.i. space having non-trivial Boyd indeces.

The sequence of rearrangements $\pi = \{\pi_n\}$

$$\pi_n(k) = \begin{cases} 2, & k = 1,\ n > 0; \\ 1, & k = 2; \\ k, & \text{for other } (n,k) \in \Omega; \end{cases}$$

satisfies the conditions of Corollaries 11.7 and 11.8. Since

$$\|\sum_{n=1}^{m} \chi_n^2\|_{L_\infty} = 1, \quad \|\sum_{n=1}^{m} \chi_n^1\|_{L_\infty} = m$$

for any integer m, it follows that T_π is not bounded in L_∞ and L_1. Consequently, Corollaries 11.7 and 11.8 fail for $p = 1$ and $p = \infty$.

Clearly, any condition of type

$$| \pi_n(k) - k | \leq M_n \tag{11.15}$$

or

$$| \{k : \pi_n(k) \neq k\} | \neq M_n$$

is not sufficient for the equivalence of $\{\chi_n^k\}$ and $\{\chi_n^{\pi_n(k)}\}$ in L_p $(p \neq 2)$ if M_n tends to ∞. There do not exist any necessary conditions of such type. For example, let $\pi_n(k) = 2^n + 1 - k$, $n \geq 1$. Then T_π coinsides with the operator $x(t) \to x(1-t)$ on the set $\{\chi_n^k\}$, $n \geq 1$. Therefore, $\|T_\pi\|_{L_p} \leq 2$ for every $p \in [1, \infty]$.

Conditions (11.13) and (11.14) are independent, the corresponding example was constructed in [255]. Therefore there exists an element $\pi \in \Pi$ such that T_π is bounded in L_p for $p \in (1,2]$ and T_π is not bounded in L_p for $p \in (2, \infty]$.

Corollary 11.9. *Let J be the interval (1,2) or $(2,\infty)$, $\pi \in \Pi$. If T_π is bounded in L_p for some $p \in J$, then T_π is bounded in L_p for any $p \in J$.*

Indeed, conditions (11.13) and (11.14) do not depend on $p \in J$. Analogously, if T_π is an isomorphism L_p for some $p \in (1,\infty)$, $p \neq 2$, then the same is true for every $p \in (1, \infty)$.

Now we shall investigate the problem of boundedness of T_π from L_q into L_q, where $1 < q < p < \infty$. The case $q \leq 2 \leq p$ is trivial. Since

$$\|T_\pi\|_{p,q} = \|T_{\pi^{-1}}\|_{q',p'},$$

we shall consider only the case $2 < q < p < \infty$. An element $\pi \in \Pi$ generates the functions

$$\begin{aligned} \mu(\tau) &= \sup_{mes\, e \leq \tau,\ e \in \mathfrak{N}} mes\, \pi(e) \\ \nu(\tau) &= \sup_{mes\, e \leq \tau,\ e \in \mathfrak{N}} mes\, \pi^{-1}(e), \end{aligned} \tag{11.16}$$

where $\pi(e)$ and $\pi^{-1}(e)$ are defined by (11.3) and (11.4).

Theorem 11.10. *Let $1 < q < p < 2$ or $2 < q < p < \infty$. An operator T_π is bounded from L_q into L_q iff*

$$\frac{\mu(t)}{t} \in L_r, \qquad 1 < q < p < 2,$$

$$\frac{\nu(t)}{t} \in L_r, \qquad 2 < q < p < \infty,$$

where

$$r = \begin{cases} \dfrac{\frac{1}{q} - \frac{1}{2}}{\frac{1}{q} - \frac{1}{p}}, & 1 < q < p < 2 \\[2ex] \dfrac{\frac{1}{2} - \frac{1}{p}}{\frac{1}{q} - \frac{1}{p}}, & 2 < q < p < \infty. \end{cases} \tag{11.17}$$

The complicated proof of this theorem will be divided into some Lemmas. Some of them are simple. Their proofs are presented for the completeness of the exposition.

Lemma 11.11. *Let $1 \leq p < \infty$, $x_i \in L_p$, $x_i \geq 0$, $\|x_i\|_{L_1} \geq \lambda_i$, $mes\,(supp\, x_i) \leq \alpha_i$ for every $i = 1, 2, \ldots$ Then*

$$\|\sum_i x_i\|_{L_p} \geq (\sum_i \lambda_i^p \alpha_i^{1-p})^{1/p}. \tag{11.18}$$

Proof. We use the semiordering $\prec$ (see Section 1.e.). If x_i and y_i are equimeasurable for every $i = 1, 2, ...$, and $\{y_i\}$ are disjoint, then

$$\sum_i x_i \succ \sum y_i.$$

By Theorem 1.e.5,

$$\|\sum_i x_i\|_{L_p} \geq \|\sum_i y_i\|_{L_p}.$$

Therefore without loss of generality we may assume that $\{x_i\}$ are disjoint and $x_i \geq 0$. Put $e_i = \operatorname{supp} x_i$ and

$$Px(t) = \sum_i \frac{1}{mes\, e_i} \int_{e_i} x(s)\, ds \kappa_{e_i}(t).$$

By Theorem 1.e.6, $\|P\|_{L_p} = 1$. Hence,

$$\begin{aligned} &\|\textstyle\sum_i x_i\|_{L_p} \geq \|P(\sum_i x_i)\|_{L_p} = \|\sum_i \frac{1}{mes\, e_i} \int_{e_i} x_i(s)\, ds \kappa_{e_i}\|_{L_p} = \\ &\|\textstyle\sum_i \frac{\lambda_i}{mes\, e_i} \kappa_{e_i}\|_{L_p} = \left(\sum_i \lambda_i^p \alpha_i^{1-p}\right)^{1/p} \end{aligned} \tag{11.19}$$

□

Clearly, Lemma 11.11 is valid for $p = \infty$ too. In that case (11.18) has the following form

$$\|\sum_i x_i\|_{L_\infty} \geq \sup_i \frac{\lambda_i}{\alpha_i}.$$

Now we present the result in the opposite direction.

Lemma 11.12. *Let* $1 \leq p < \infty$, $0 < \alpha < 1$, $a_k \geq 0$, $e_k \subset [0,1]$, $mes\, e_{k+1} \leq \alpha\, mes\, e_k$ *for each* $k = 1, 2, ...$ *There exists a constant* $C(\alpha, p) > 0$ *such that*

$$\|\sum_{k=1}^{\infty} a_k \kappa_{e_k}\|_{L_p} \leq C(\alpha, p) (\sum_{k=1}^{\infty} a_k^p mes\, e_k)^{1/p}. \tag{11.20}$$

Proof. Let $x_k \in L_1$, $k = 1, 2, ...$ By (1.c.2),

$$\sum_{k=1}^{\infty} x_k \prec \sum_{k=1}^{\infty} x_k^*.$$

Therefore

$$\|\sum_{k=1}^{\infty} a_k \kappa_{e_k}\|_{L_p} \leq C(\alpha, p) \|\sum_{k=1}^{\infty} a_k \kappa_{[0, mes\, e_k]}\|_{L_p}. \tag{11.21}$$

Consider the function

$$z(t)=\sum_{k=1}^{\infty} a_k\kappa_{(\gamma_{k+1},\gamma_k)}(t),$$

where

$$\gamma_k=\sum_{i=k}^{\infty} mes\, e_i.$$

It is well known that the operator

$$H_1x(t)=\int_t^1 \frac{x(s)}{s}\,ds$$

is bounded in L_p, $1\le p<\infty$. It is easy to check that

$$H_1z(\gamma_n)=\int_{\gamma_n}^1 \frac{z(s)}{s}\,ds=\sum_{k=1}^{n-1} a_k\int_{\gamma_{k+1}}^{\gamma_k}\frac{ds}{s}=\sum_{k=1}^{n-1} a_k\ln\frac{\gamma_k}{\gamma_{k+1}}.$$

Since

$$\gamma_k=\sum_{i=k}^{\infty} mes\, e_i\ge\frac{1}{\alpha}\sum_{i=k}^{\infty} mes\, e_{i+1}=\frac{1}{\alpha}\gamma_{k+1},$$

then

$$H_1z(\gamma_n)\ge\ln\frac{1}{\alpha}\sum_{k=1}^{n-1} a_k$$

for every $n=1,2,...$ The function $H_1z(t)$ decreases on $(0,1]$. Therefore

$$H_1z(t)\ge\ln\frac{1}{\alpha}\sum_{n=2}^{\infty}\Big(\sum_{k=1}^{n-1} a_k\Big)\kappa_{(\gamma_{n+1},\gamma_n)}(t)$$

and

$$\|H_1z+\ln\tfrac{1}{\alpha}z\|_{L_p}\ge\ln\tfrac{1}{\alpha}\|\textstyle\sum_{n=1}^{\infty}(\sum_{k=1}^{n} a_k)\kappa_{(\gamma_{n+1},\gamma_n)}\|_{L_p}=$$

$$=\ln\tfrac{1}{\alpha}(\textstyle\sum_{n=1}^{\infty}(\sum_{k=1}^{n} a_k)^p(\gamma_n-\gamma_{n+1}))^{1/p}=$$

$$=\ln\tfrac{1}{\alpha}(\textstyle\sum_{n=1}^{\infty}(\sum_{k=1}^{n} a_k)^p mes\, e_n)^{1/p}.$$

Hence,

$$\|\textstyle\sum_{k=1}^{\infty} a_k\kappa_{(0,mes\, e_k)}\|_{L_p}\le(\sum_{n=1}^{\infty}(\sum_{k=1}^{n} a_k)^p mes\, e_n)^{1/p}\le\frac{1}{\ln\frac{1}{\alpha}}\|H_1z+$$

$$+\ln\tfrac{1}{\alpha}z\|_{L_p}\le(\|H_1\|_p\ln^{-1}\tfrac{1}{\alpha}+1)\|z\|_{L_p}=$$

$$=(\|H_1\|_p\ln^{-1}\tfrac{1}{\alpha}+1)(\textstyle\sum_{k=1}^{\infty} a_k^p mes\,_k)^{1/p}.$$

Comparing the obtained inequality with (11.21), we get (11.20), where $C(\alpha,p) = \|H_1\|_p \ln^{-1}\frac{1}{\alpha} + 1$. □

Clearly, Lemma 11.12 fails if $p = \infty$.

Let us consider the positively homogeneous operator

$$J\,x(t) = \sum_{j=1}^{\infty} 2^j \int_0^{2^{-j}} (x^*(s) - x^*(2^{-j}))\, ds\, \kappa_{(0,2^{-j})}(t) + \int_0^1 |\, x(s)\,|\; ds.$$

Lemma 11.13. *For any $x \in L_1$, $\quad x \prec 4\,J\,x$.*

Proof. The function $J\,x(t)$ does not increase on $(0,1]$ and $x^*(s) \geq x^*(2^{-j-1})$ for $s \in (0, 2^{-j-1})$. As a consequence we have

$$\int_0^{2^{-n}} (J\,x)^*(t)\,dt = \int_0^{2^{-n}} Jx(t)\,dt \geq \sum_{j=1}^{n} 2^{j-n} \int_0^{2^{-j}} (x^*(s) -$$

$$-x^*(2^{-j}))\,ds + 2^{-n} \int_0^1 |\, x(s)\,|\; ds \geq$$

$$\geq \sum_{j=1}^{n-1} 2^{j-n} 2^{-j-1} (x^*(2^{-j-1}) - x^*(2^{-j})) +$$

$$+ \int_0^{2^{-n}} x^*(s)\,ds - 2^{-n} x^*(2^{-n}) +$$

$$+2^{-n} \int_0^1 |\, x(s)\,ds = 2^{-n-1}(x^*(2^{-n}) -$$

$$-x^*(2^{-1})) - 2x^*(2^{-n}) + 2\int_0^1 |\, x(s)\,|\; ds) +$$

$$+ \int_0^{2^{-n}} x^*(s)\,ds \geq -2^{-n-1} x^*(2^{-n}) +$$

$$+ \int_0^{2^{-n}} x^*(s)\,ds \geq \frac{1}{2} \int_0^{2^{-n}} x^*(s)\,ds.$$

If $\tau \in (2^{-n}, 2^{-n+1})$ for some $n \in \{1, 2, \ldots\}$, then

$$\int_0^{\tau} J\,x(t)\,dt \geq \int_0^{2^{-n}} J\,x(t)\,dt \geq \frac{1}{2} \int_0^{2^{-n}} x^*(s)\,ds \geq$$

$$\geq \frac{1}{4} \int_0^{2^{-n+1}} x^*(s)\,ds \geq \frac{1}{4} \int_0^{\tau} x^*(s)\,ds.$$

□

Each Haar polynomial

$$x(t) = \sum_{(n,k)\in\Omega} c_{n,k} \chi_n^k(t)$$

generates the function

$$y(t) = \sum_{(n,k)\in\Omega} c_{n,k}^2 \kappa_{\Delta_n^k}(t).$$

Lemma 11.14. *Let x be a Haar polynomial, $\|x\|_{L_\infty} \le 1$, $0 \le \tau \le 1$, and $\pi \in \Pi$. Then*

$$\int_0^\tau ((S_\pi y)^*(t) - (S_\pi y)^*(\tau))\, dt \le 4\nu(\tau),$$

where the function ν is defined in (11.16).

Proof. By definition,

$$S_\pi y(t) = \sum_{(n,k)\in\Omega} c_{n,k}^2 \kappa_{\Delta_n}^{\pi_n(k)(t)}.$$

Given $\tau \in (0,1]$, we can find a decomposition

$$c_{n,k}^2 = u_{n,k}^2 + v_{n,k}^2, \quad (n,k) \in \Omega,$$

such that

$$\sum_{(n,k)\in\Omega} u_{n,k}^2 \kappa_{\Delta_n^{\pi_n(k)}}(t) = \min(S_\pi y(t), (S_\pi y)^*(\tau))$$

and

$$\sum_{(n,k)\in\Omega} v_{n,k}^2 \kappa_{\Delta_n}^{\pi_n(k)}(t) = \max(S_\pi y(t) - (S_\pi y)^*(\tau), 0). \tag{11.22}$$

Denote the function (11.22) by $v(t)$ and its support by e. Clearly, $mes\, e \le \tau$ and

$$\int_0^\tau ((S_\pi y)^*(t) - (S_\pi y)^*(\tau))\, dt = \int_e v(t)\, dt. \tag{11.23}$$

By definition of the function ν, $mes\, \pi^{-1}(e) \le \nu(\tau)$. We have

$$\int_e v(t)\, dt = \int_0^1 \sum_{(n,k)\in\Omega} v_{n,k}^2 \kappa_{\Delta_n}^k(t)\, dt = \int_0^1 (\sum_{(n,k)\in\Omega} v_{n,k}\chi_n^k(t))^2\, dt. \tag{11.24}$$

Denote

$$\sigma = \{(n,k) : (n,k) \in \Omega, \quad v_{n,k} \neq 0\}.$$

By the Parseval equality,

$$\int_0^1 (\sum_{(n,k)\in\Omega} v_{n,k}\chi_n^k(t))^2\, dt \le \int_0^1 (\sum_{(n,k)\in\Omega} c_{n,k}\chi_n^k(t))^2\, dt. \tag{11.25}$$

Since

$$\sum_{(n,k)\in\Omega} v_{n,k}\kappa_{\Delta_n^k}(t) = 0$$

for $t \notin \pi^{-1}(e)$, then

$$\sum c_{n,k}\chi_n^k(t) = 0$$

for $t \notin \pi^{-1}(e)$. We may assume that σ has the following property. If $(n,k) \notin \sigma$, $u_{n,k} > 0$ and $(n,k) \in O(i,j)$, then $(i,j) \notin \sigma$. Therefore

$$\sum_{(n,k)\notin\sigma} c_{n,k}\chi_n^k(t)$$

is the conditional expectation of $x(t)$ with respect to σ-algebra $\{\Delta_n^k,\ (n,k) \notin \sigma, \quad u_{n,k} > 0\}$. Hence,

$$\mid \sum_{(n,k)\notin\sigma} c_{n,k}\chi_n^k(t) \mid\leq 1$$

and

$$\mid \sum_{(n,k)\in\sigma} c_{n,k}\chi_n^k(t) \mid\leq 2.$$

Consequently,

$$\int_0^1 (\sum_{(n,k)\in\Omega} c_{n,k}\chi_n^k(t))^2\, dt \leq \int_{\pi^{-1}(e)} 4\, dl \leq 4\nu(\tau). \tag{11.26}$$

Combining (11.23-26), we get

$$\int_0^\tau ((S_\pi y)^*(t) - (S_\pi y)^*(\tau))\, dt \leq 4\nu(\tau). \quad \square$$

Proof of Theorem 11.10. We consider only the case $2 < q < p < \infty$. The other case is dual. Without loss of generality we may assume that there exist $\tau_k \in (0,1]$ such that $\nu(\tau_k) = 2^{-k}$, $\quad k = 0,1,2,\ldots$ Given $\varepsilon \in (0, \frac{1}{2})$, we find $e_k, g_k \in \mathfrak{N}$ such that $mes\, e_k \leq \tau_k$, $\quad (1-\varepsilon)2^{-k} \leq mes\, g_k \leq 2^{-k}$ and $\operatorname{supp} S_\pi \kappa_{g_k} \subset e_k$, $\quad k = 1,2,\ldots$ Given an integer N, we consider the function

$$x_N(t) = \sum_{k=1}^N x_k \kappa_{g_k}(t),$$

where $x_k \geq 0$ will be chosen later. By Lemma 11.12 there exists a constant $C_1 > 0$ such that

$$\|x_N\|_{L_{\frac{p}{2}}} \leq C_1(\sum_{k=1}^N x_k^{\frac{p}{2}} 2^{-k})^{\frac{2}{p}}.$$

We have

$$\int_0^1 S_\pi \kappa_{g_k}(t)\, dt = \int_0^1 \kappa_{g_k}(t)\, dt = mes\, g_k \geq (1-\varepsilon)2^{-k}$$

and

$$mes\,(\operatorname{supp} S_\pi \kappa_{g_k}) \leq mes\, e_k \leq \tau_k$$

for any $k = 1, 2, \ldots$ Applying Lemma 11.11, we get

$$\|S_\pi x_N\|_{L_{\frac{q}{2}}} \geq (1-\varepsilon) \left(\sum_{k=1}^{n} \left(\frac{x_k 2^{-k}}{\tau_k}\right)^{\frac{q}{2}} \tau_k \right)^{\frac{2}{q}}$$

Hence,

$$\|S_\pi\|_{\frac{p}{2},\frac{q}{2}} \geq \tfrac{1-\varepsilon}{C_1} \sup_{x_n} \frac{\left(\sum_{k=1}^N x_k^{\frac{q}{2}} 2^{-\frac{kq}{2}} \tau_k^{1-\frac{q}{2}}\right)^{\frac{2}{q}}}{\left(\sum_{k=1}^N x_k^{\frac{p}{2}} 2^{-k}\right)^{\frac{2}{p}}} =$$

$$= \tfrac{1-\varepsilon}{C_1} \left(\sum_{k=1}^N 2^{-\frac{kq}{2}(1-\frac{2}{p})\frac{p}{p-q}} \tau_k^{\frac{(2-q)p}{2(p-q)}}\right)^{\frac{2(p-q)}{pq}}.$$

Since

$$\frac{q}{2}(1-\frac{2}{p})\frac{p}{p-q} = r, \qquad \frac{(2-q)p}{2(p-q)} = 1-r,$$

then

$$\|S_\pi\|_{\frac{p}{2},\frac{q}{2}} \geq \frac{1-\varepsilon}{C_1} \left(\sum_{k=1}^{N} 2^{-kr} \tau_k^{1-r}\right)^{\frac{2(p-q)}{pq}}.$$

We have

$$\int_0^1 \left(\frac{\nu(\tau)}{\tau}\right)^r d\tau \leq \sum_{k=1}^{\infty} \nu^r(\tau_{k-1}) \int_{\tau_k}^{\tau_{k-1}} \tau^{-r}\, d\tau \leq \sum_{k=1}^{\infty} 2^{-(k-1)r} \frac{\tau_k^{1-r}}{r-1}.$$

The last estimates show that

$$\int_0^1 \left(\frac{\nu(\tau)}{\tau}\right)^r d\tau < \infty.$$

Sufficiency. Let

$$x(t) = \sum_{(n,k)\in\Omega} x_{n,k} \chi_n^k(t)$$

be a Haar polynomial and $\|x\|_{L_\infty} \leq 1$. Put

$$y(t) = \sum_{(n,k)\in\Omega} x_{n,k}^2 \kappa_{\Delta_n^k}(t).$$

Using Lemmas 11.13 and 11.14, we get

$$\tfrac{1}{4} S_\pi y \prec J\, S_\pi y = \textstyle\sum_{j=1}^\infty 2^j \int_0^{2^{-j}} \left((S_\pi y)^*(s) - (S_\pi y)^*(2^{-j})\right) ds \kappa_{(0,2^{-j})}(t) +$$

$$+ \textstyle\int_0^1 S_\pi y(s)\, ds \leq 4 \sum_{j=1}^\infty 2^j \nu(2^{-j}) \kappa_{(0,2^{-j})}(t) + 1 =$$

$$= 4 \textstyle\sum_{j=1}^\infty \left(\sum_{k=1}^j 2^k \nu(2^{-k})\right) \kappa_{(2^{-j-1},2^{-j})}(t) + 1.$$

The function $\nu(t)$ does not decrease on $[0,1]$. Therefore, for $t \in (2^{-j-1}, 2^{-j})$, we have

$$\int_t^1 \frac{\nu(\tau)}{\tau^2}\, d\tau \geq \sum_{k=1}^j \int_{2^{-k}}^{2^{-k+1}} \frac{\nu(\tau)}{\tau^2}\, d\tau \geq$$

$$\geq \sum_{k=1}^j \nu(2^{-k}) \int_{2^{-k}}^{2^{-k+1}} \frac{d\tau}{\tau^2} = \sum_{k=1}^j \nu(2^{-k}) 2^{-k+1}.$$

Hence,

$$\frac{1}{4} S_\pi y(t) \prec 8 \int_t^1 \frac{\nu(\tau)}{\tau^2}\, d\tau + 1 = 8H_1(\frac{\nu(\tau)}{\tau})(t) + 1.$$

By 1.e.5,

$$\|S_\pi y\|_{L_r} \leq 32 \|H_1\|_r \|\frac{\nu(\tau)}{\tau}\|_{L_r} + 4.$$

The obtained inequality shows that T_π is bounded from L_∞ into L_{2r}. T_π is isometric in L_2. The M. Riesz ionterpolation theorem implies the boundedness of T_π from L_p into L_q. □

This theorem leads to simple sufficient conditions of the boundedness of T_π. Given a sequence $\pi = (\pi_0, \pi_1, \ldots)$, we put

$$\omega_n = \operatorname{card}\{k : \pi_n(k) \neq k\}.$$

If x_n tends to 0, then we denote by x_n^* the decreasing rearrangement of $|x_n|$.

Corollary 11.15. *Let numbers p, q, r satisfy condition (11.17). If*

$$\sum_{n=1}^\infty (\omega_n 2^{-n})^* n^{r-1} < \infty,$$

then T_π is bounded from L_p into L_q.

Corollary 11.16. *Let numbers p, q, r satisfy condition (11.17) and*

$$\sum_{n=1}^\infty (\max_{1 \leq k \leq 2^n} | \pi_n(k) - k | 2^{-n})^* n^{r-1} < \infty.$$

Then T_π acts from L_p into L_q.

To prove these statements it is sufficient to verify the validity of Theorem 11.10 conditions. This may be found in [255].

Operators T_π have some interesting properties. Given a measurable subset $e \subset [0,1]$, denote by Q_e the operator of multiplication by a characteristic function of e. Let m_n be a sequence of integers such that $\{m_n 2^{-n}\}$ is dense on $[0,1]$. Put

$$\bar{\pi}_n(k) = (k + m_n) \bmod 2^n + 1. \tag{11.27}$$

Then $T_{\bar{\pi}}$ has the following properties.

1. Let E be an r.i. space and $T_{\bar{\pi}}$ is bounded in E. Then $E = L_2$ up to equivalence.

2. If $mes\, e > 0$, $mes\, f > 0$, then

$$\|Q_e T_{\bar{\pi}} Q_f\|_{L_2} = 1.$$

We present another example. Let $\{i_n\}$, $\{j_n\}$ be sequences of integers such that the set $(i_n 2^{-n}, j_n 2^{-n})$ is dense in the unit square. We put

$$\tilde{\pi}_n(k) = \begin{cases} j_n, & k = i_n; \\ i_n, & k = j_n \quad (n = 0, 1, \ldots); \\ k, & k \neq i_n, j_n. \end{cases}$$

Then $T_{\tilde{\pi}}$ is bounded in L_p for every $p \in (1, \infty)$ and

$$\|Q_{(\alpha,\beta)} T_{\tilde{\pi}} Q_{(\gamma,\delta)}\|_p \geq 1$$

for every $0 \leq \alpha < \beta \leq 1$, $0 \leq \gamma < \delta \leq 1$, $p \in (1, \infty)$.

The characteristics of the sequence $\pi \in \Pi$ which were introduced at the beginning of this section prove to be useful for the investigation of the Haar rearrangements in the space BMO. Here we shall apply the single-index H.s.

Recall, if

$$x(t) = \sum_n c_n \chi_n(t),$$

then

$$\|x\|_{BMO} = \sup_{e \in \mathfrak{N}} \Big(\frac{1}{mes\, e} \sum_{\Delta_n \subset e} c_n^2 mes\, \Delta_n\Big)^{\frac{1}{2}},$$

where $\mathfrak{N}$ is the set of the finite union of disjoint dyadic intervals. The designation

$$\sup_{e \in \mathfrak{N}} \frac{mes\, \pi^{-1}(e)}{mes\, e} = \|\pi^{-1}\|$$

was introduced in [264].

Theorem 11.17. *Let $\pi \in \Pi$. Then*

$$\|T_\pi\|_{BMO} = \|\pi^{-1}\|^{\frac{1}{2}}. \tag{11.28}$$

Proof. Given $\varepsilon > 0$, we can find $e_0 \in \mathfrak{N}$ and a set of disjoint dyadic intervals $\Delta_{n_1}, \ldots, \Delta_{n_s}$ such that $\Delta_{\pi(i)} \subset e_0$ for every $i = 0, 1, 2, \ldots, s$ and

$$\frac{\sum_{i=1}^{s} mes\, \Delta_{n_i}}{mes\, e_0} \geq \|\pi^{-1}\| - \varepsilon.$$

Put

$$x_0(t) = \sum_i \chi_{n_i}(t).$$

Then

$$\|x_0\|_{BMO} = 1.$$

and

$$\|T_\pi x_0\|_{BMO} = \| \textstyle\sum_{i=1}^{s} \chi_{\pi(n_i)}\|_{BMO} \geq (\frac{1}{mes\, e_0} \sum_{\Delta_{\pi(n)} \subset e_0} mes\, \Delta_{\pi(n)})^{\frac{1}{2}} =$$

$$= (\frac{\sum_{i=1}^{s} mes\, \Delta_{n_i}}{mes\, e_0})^{\frac{1}{2}} \geq (\|\pi^{-1}\| - \varepsilon)^{\frac{1}{2}}.$$

Hence,

$$\|T_\pi\|_{BMO} \geq \|\pi^{-1}\|^{\frac{1}{2}}. \tag{11.29}$$

Now we prove the opposite inequality. Let $\varepsilon > 0$ and

$$x(t) = \sum_{i=1}^{s} c_i \chi_i(t)$$

is a Haar polynomial. We can find $e_0 \in \mathfrak{N}$ such that

$$\|\sum_{i=1}^{s} c_i \chi_{\pi(i)}\|_{BMO} \leq (\frac{1}{mes\, e_0} \sum_{\Delta_i \subset e_0} c_{\pi(i)}^2 mes\, \Delta_i)^{\frac{1}{2}} + \varepsilon.$$

Then

$$\|T_\pi x\|_{BMO} \leq (\frac{1}{mes\, e_0} \textstyle\sum_{\Delta_i \subset e_0} c_{\pi(i)}^2 mes\, \Delta_i)^{\frac{1}{2}} + \varepsilon =$$

$$= \left(\frac{mes\, \pi^{-1}(e_0)}{mes\, e_0}\right)^{\frac{1}{2}} \left(\frac{1}{mes\, \pi^{-1}(e_0)} \textstyle\sum_{\Delta_i \subset e_0} c_{\pi(i)}^2 mes\, \Delta_{\pi(i)}\right)^{\frac{1}{2}} + \varepsilon \leq$$

$$\leq \|\pi^{-1}\|^{\frac{1}{2}} \|x\|_{BMO} + \varepsilon.$$

The obtained inequality and (11.29) prove (11.28). □

Since $[BMO, L_2]_\Theta = L_{\frac{2}{\Theta}}$, $0 < \Theta < 1$ [21], then Theorem 11.17 leads to a more simple proof of the sufficiency of Theorem 11.5.

Chapter 12

Fourier-Haar Multipliers

In this section we investigate multipliers with respect to the H.s. A sequence $\lambda_{n,k}$, $(n, k) \in \Omega$ generates the operator

$$\Lambda\Big(\sum_{(n,k)\in\Omega} a_{n,k}\chi_n^k\Big) = \sum \lambda_{n,k} c_{n,k}\chi_n^k$$

on the polynomials with respect to the H.s. Such operators are said to be multipliers. Recall that the norm of Λ from L_p into L_q (L_p) is denoted by $\|\Lambda\|_{p,q}$ $(\|\Lambda\|_p)$. The main result of Chapter 5 (Corollary 5.8) may be formulated in the following way. If $1 < p < \infty$, $\frac{1}{p} + \frac{1}{p'} = 1$, then

$$\|\Lambda\|_p \le (\max(p, p') - 1) \sup_{(n,k)\in\Omega} |\lambda_{n,k}| .$$

Theorem 12.1. *Let* $1 < p \le q < \infty$. *Then*

$$\|\Lambda\|_{p,q} \approx \sup_{(n,k)\in\Omega} 2^{n(\frac{1}{p}-\frac{1}{q})} |\lambda_{n,k}|,$$

where the equivalence constants depend only on p, q.

Proof. We can suppose that $p < q$. Denote $\gamma = \frac{1}{p} - \frac{1}{q}$.

Let $|\lambda_{n,k}| \le 2^{-n\gamma}$ for every $(n, k) \in \Omega$. Standard reasoning shows that Λ is an integral operator, its kernel is the function

$$K(t, s) = \sum_{(n,k)\in\Omega} \lambda_{n,k} 2^n \chi_n^k(t)\chi_n^k(s).$$

Let m be an integer and $2^{-m} \le |t - s| \le 2^{-m+1}$. Then

$$K(t,s) \le \sum_{(n,k)\in\Omega} 2^n |\lambda_{n,k}\chi_n^k(t)\chi_n^k(s)| \le$$

$$\le \sum_{n=0}^{m} 2^{-n\gamma+n} \le \frac{2^{-m\gamma+m+1}}{1-2^{\gamma-1}}.$$

Hence,

$$K(t,s) \leq \frac{4}{1-2^{\gamma-1}} \mid t-s \mid .$$

By the classical Hardy-Littlewood theorem, the operator

$$\int_0^1 \frac{x(s)}{\mid t-s \mid^{1-\gamma}} \, ds$$

acts from L_p into L_q if $1 < p < q < \infty$, $\frac{1}{p} - \frac{1}{q} = \gamma$. Consequently,

$$\|\Lambda\|_{p,q} \leq C(p,q). \tag{12.1}$$

On the other hand, for every $(n,k) \in \Omega$,

$$\|\Lambda\|_{p,q} \geq \frac{\|\Lambda \chi_n^k\|_{L_q}}{\|\chi_n^k\|_{L_p}} = \mid \lambda_{n,k} \mid 2^{n(\frac{1}{p}-\frac{1}{q})}.$$

Hence,

$$\|\Lambda\|_{p,q} \geq \sup_{(n,k)\in\Omega} 2^{n(\frac{1}{p}-\frac{1}{q})} \mid \lambda_{n,k} \mid . \quad \square$$

Theorem 12.1 fails for $p = 1$ or $q = \infty$. Indeed, the function $\sum_{n=1}^{\infty} 2^{\frac{n}{p}} \frac{1}{n} \chi_n^1(t)$ belongs to L_p, $1 < p < \infty$, but the function $\sum_{n=1}^{\infty} \frac{1}{n} \chi_n^1(t)$ is not bounded. This shows that the multiplier $\lambda_{n,k} = 2^{-\frac{n}{p}}$ for $(n,k) \in \Omega$ does not act from L_p into L_∞. By duality argument it is not bounded from L_1 into $L_{p'}$.

Let $\lambda_{n,k} = 2^{-n}$, $(n,k) \in \Omega$ and

$$x_N(t) = \sum_{n=0}^{N} 2^n \chi_n^1, \quad N = 1, 2, \ldots$$

Then $\|x_N\|_{L_1} < 2$ and

$$\|\Lambda x_N\|_{L_\infty} = \|\sum_{n=0}^{N} \chi_n^1\|_{L_\infty} = N + 1.$$

Therefore this multiplier does not act from L_1 into L_∞.

Sometimes we shall write λ_m instead of $\lambda_{n,k}$, where $m = 2^n + k$, $(n,k) \in \Omega$. Given a sequence $\{\lambda_n\}$, we construct the function

$$\mu(t) = \sup_{n \geq 1} \mid \lambda_n \chi_n(t) \mid . \tag{12.2}$$

Theorem 12.2. *Let $1 \leq q < p \leq \infty$, $(p,q) \neq (\infty, 1)$, $\frac{1}{r} = \frac{1}{q} - \frac{1}{p}$. The multiplier Λ is bounded from L_p into L_q iff $\mu \in L_r$. Moreover,*

$$\|\Lambda\|_{p,q} \approx \|\mu\|_{L_r}, \tag{12.3}$$

where the equivalence constants depend only on r.

Proof. Necessity . At first we construct a set $P \subset \mathbf{N}$ such that

1) $\mu(t) \leq 2\sup_{n\in P} \mid \lambda_n \chi_n(t) \mid$;

2) if $n_1, n_2 \in P$ and $\Delta_{n_1} \subset \Delta_{n_2}$, then $|\lambda_{n_2}| \geq 2|\lambda_{n_1}|$.

To this end we introduce a semiordering relation on $\mathbf{N}$. Put $i \prec j$ if $\Delta_i \supset \Delta_j$. For simplicity we assume that $\mid \lambda_1 \mid + \mid \lambda_2 \mid > 0$. We construct P by induction.

If $\mid \lambda_1 \mid > \mid \lambda_2 \mid$ $(\mid \lambda_1 \mid \leq \mid \lambda_2 \mid)$, then $1\ (2) \in P$. Suppose that $n \in \mathbf{N}$, $n > 1$, and the set $P \cap \{1, 2, ..., n-1\}$ is constructed. Define

$$m = \max\{k : 1 \leq k < n,\ k \in P,\ k \prec n\}.$$

Then $n \in P$ iff $\mid \lambda_n \mid \geq 2 \mid \lambda_m \mid$.

Clearly, property 2) is satisfied. If $j \in \mathbf{N}$, $j \notin P$, then there exists an $i \in P$ such that $i \prec j$ and $\mid \lambda_j \mid < 2 \mid \lambda_i \mid$. Therefore property 1) is satisfied too.

Put

$$\nu(t) = \sup_{n\in P} \mid \lambda_n \chi_n(t). \tag{12.4}$$

It is evident that, for every $s > 0$,

$$\nu^s(t) \leq \sum_{n\in P} \mid \lambda_n \chi_n(t) \mid^s . \tag{12.5}$$

Given $t_0 \in (0,1)$, denote $\nu(t_0) = \alpha$. There exists a sequence of dyadic intervals $\Delta_{n_i} (1 \leq i \leq j)$ such that

$$t_0 \in \Delta_{n_j} \subset \Delta_{n_{j-1}} \subset ...\Delta_{n_1}$$

and $\mid \lambda_{n_j} \mid = \alpha$. By property 2),

$$\mid \lambda_{n_k} \mid \leq \frac{1}{2} \mid \lambda_{n_{k+1}} \mid$$

for every $k = 1, 2, ..., j-1$. Hence,

$$\sum_{k=1}^{j} \mid \lambda_{n_k} \mid^s \leq \sum_{k=0}^{\infty} 2^{-ks}\alpha^s = \frac{\alpha^s}{1-2^{-s}}.$$

It means that, for every $t \in [0,1)$,

$$\sum_{n\in P} \mid \lambda_n \chi_n(t) \mid^s \leq \frac{1}{1-2^{-s}} \nu^s(t). \tag{12.6}$$

So,

$$\nu^s(t) \leq \sum_{n\in P} \mid \lambda_n \chi_n(t) \mid^s \leq \frac{1}{1-2^{-s}} \nu^s(t)$$

for every $t \in [0,1]$, $s > 0$.

Let Λ be a bounded operator from L_p into L_q. Then Λ acts from $L_{q'}$ into $L_{p'}$, where p', q' are the conjugate exponents. By the M. Riesz interpolation theorem (1.e.2), Λ is bounded from L_{p_0} into L_{q_0}, and

$$\|\Lambda\|_{p_0,q_0} \leq \|\Lambda\|_{p,q} = \|\Lambda\|_{q',p'},$$

where

$$p_0 = \frac{2r}{r-1}, \quad q_0 = \frac{2r}{r+1}.$$

Given a sequence $\varepsilon_n = \pm 1$, we put

$$x_\varepsilon(t) = \sum_{n \in P} \varepsilon_n \mid \lambda_n \mid^{\frac{r-1}{2}} \chi_n(t).$$

Applying (12.6), we get

$$\mid x_\varepsilon(t) \mid \leq \sum_{n\in P} \mid \lambda_n \chi_n(t) \mid^{\frac{r-1}{2}} \leq (1 - 2^{\frac{1-r}{2}})^{-1} \nu^{\frac{r-1}{2}}(t).$$

Therefore

$$\begin{aligned} \|x_\varepsilon\|_{L_{p_0}} &\leq (1-2^{\frac{1-r}{2}})^{-1} \left(\int_0^1 \nu^{\frac{r-1}{2}\cdot\frac{2r}{r-1}}(t)\, dt\right)^{\frac{r-1}{2r}} = \\ &= (1-2^{\frac{1-r}{2}})^{-1} \|\nu\|_{L_r}^{\frac{r-1}{2}}. \end{aligned} \tag{12.7}$$

By Corollary 6.4 and (12.5), we have

$$\begin{aligned} &\sup_{\varepsilon_n=\pm1} \|\Lambda x_\varepsilon\|_{L_{q_0}} = \sup_{\varepsilon_n=\pm1} \|\textstyle\sum_{n\in P} \varepsilon_n |\lambda_n|^{\frac{r+1}{2}} \chi_n\|_{L_{q_0}} \geq \\ &\geq \tfrac{1}{\sqrt{2}} \left(\int_0^1 \left(\textstyle\sum_{n\in P} (\mid \lambda_n \mid^{\frac{r+1}{2}} \chi_n(t))^2\right)^{\frac{r}{r+1}} dt\right)^{\frac{r+1}{2r}} \geq \\ &\geq \tfrac{1}{\sqrt{2}} \|\nu\|_{L_r}^{\frac{r+1}{2}}. \end{aligned} \tag{12.8}$$

It follows from (12.7) and (12.8) that

$$\|\Lambda\|_{p_0,q_0} \geq 2^{-\frac{1}{2}}(1-2^{\frac{1-r}{2}})\|\nu\|_{L_r}.$$

Consequently, the inequality

$$\|\Lambda\|_{p_0,q_0} \geq 2^{-\frac{1}{2}}(1-2^{\frac{1-r}{2}})\|\nu\|_{L_r} \geq 2^{-\frac{3}{2}}(1-2^{\frac{1-r}{2}})\|\mu\|_{L_r}$$

holds for every p, q such that $\frac{1}{q} - \frac{1}{p} = \frac{1}{r}$.

Sufficiency. We suppose that $1 < r < \infty$. Therefore $1 < r' < \infty$. Let $x(t) = \sum_{n=1}^{\infty} a_n \chi_n(t)$. Using Theorem 6.3 and the Holder inequality, we get

$$\|\Lambda x\|_{L_1} = \| \sum_{n=1}^{\infty} \lambda_n a_n \chi_n \|_{L_1} \leq c \| \left(\sum_{n=1}^{\infty} (\lambda_n a_n \chi_n)^2\right)^{1/2} \|_{L_1} \leq$$

$$\leq c \int_0^1 \sup_n \mid \lambda_n \chi_n(t) \mid \left(\sum_{n=1}^{\infty} (a_n \chi_n(t))^2\right)^{1/2} dt \leq$$

$$\leq c \|\mu\|_{L_r} \| \left(\sum_{n=1}^{\infty} (a_n \chi_n(t))^2\right)^{1/2} \|_{L_{r'}}.$$

Since the H.s. is an unconditional basis in $L_{r'}$ then there exists a constant $A = A(r')$ such that

$$\|(\sum_{n=1}^{\infty} (a_n \chi_n(t))^2)^{1/2}\|_{L_{r'}} \leq A \| \sum_{n=1}^{\infty} a_n \chi_n \|_{L_{r'}}.$$

Hence,

$$\|\Lambda x\|_{L_1} \leq c\, A \|\mu\|_{L_r} \|x\|_{L_{r'}}.$$

Consequently,

$$\|\Lambda\|_{L_{r'},1} \leq c\, A \|\mu\|_{L_r}.$$

Applying the duality argument and the M. Riesz interpolation theorem again, we get

$$\|\Lambda\|_{p,q} \leq c\, A \|\mu\|_{L_r}. \quad \square$$

Corollary 12.3. *Let* $1 \leq p_1, q_1, p_2, q_2 \leq \infty$ *and*

$$0 < \frac{1}{q_1} - \frac{1}{p_1} = \frac{1}{q_2} - \frac{1}{p_2} < 1.$$

If the multiplier Λ *acts from* L_{p_1} *into* L_{q_1}, *then* Λ *acts from* L_{p_2} *into* L_{q_2}.

Theorem 12.2 is partly valid for $p = 1$, $q = \infty$.

Theorem 12.4. *Let* Λ *be a multiplier and* $\mu \in L_1$. *Then* Λ *is bounded from* L_∞ *into* L_1.

Proof. Put

$$u_n = |\lambda_n|^{1/2}, \quad v_n = |\lambda_n|^{1/2} sign\, \lambda_n.$$

Clearly,

$$\sup_{n \geq 1} \mid u_n \chi_n(t) \mid = \sup_{n \geq 1} \mid v_n \chi_n(t) \mid = (\mu(t))^{1/2}.$$

The multiplier corresponding to the sequence $\{u_n\}$ ($\{v_n\}$) is denoted by U (V). By Theorem 12.2, U acts from L_2 into L_1 and V acts from L_∞ into L_2. Then $\Lambda = UV$ acts from L_∞ into L_1 and

$$\|\Lambda\|_{\infty,1} \leq C \|\mu\|_{L_1},$$

where C is an absolute constant. $\square$

Chapter 13

Pointwise Estimates of Multipliers

Let T be a linear operator acting in a pair of r.i. spaces. An operator T_1 is said to be a transposed one with respect to T if

$$\int_e T\kappa_f(t)\,dt = \int_f T_1\kappa_e(t)\,dt$$

for every measurable subsets $e, f \subset [0,1]$. A.P. Calderon proved the following theorem [52]. If T and its transposed operator T_1 have the weak types (1,1) and (2,2) with norms ≤ 1, then, for every $t \in (0,1]$

$$(Tx)^{**}(t) \le 2Sx^{**}(t),$$

where

$$x^{**}(t) = \frac{1}{t}\int_0^1 x^*(s)\,ds, \quad Sx(t) = \int_0^t \frac{x(s)}{t+s}\,ds.$$

It was proved in [127], [250] that under the above assumptions

$$(Tx)^*(t) \le CSx^*(t),$$

where C is an absolute constant ($C \le 60$).

Let Λ be a mutiplier with respect to the H.s. generated by a sequence $\{\lambda_{n,k}\}, (n,k) \in \Omega$ (cf. Chapter 12). Clearly, Λ coinsides with its transposed one. If $|\lambda_{n,k}| = 1$ for each $(n,k) \in \Omega$, then Λ is isometric in L_2 and Λ has the weak type (1,1) (cf. Chapter 5). Therefore

$$(\Lambda x)^*(t) \le C\int_0^1 \frac{x^*(s)}{t+s}\,ds \tag{13.1}$$

for every $t \in (0,1]$ and for some constent $C > 0$. One can show that estimate (13.1) is valid if $|\lambda_{n,k}| \le 1$ for every $(n,k) \in \Omega$.

We investigate the problem of exactness of (13.1). To this end we introduce the functional

$$G_t(x) = \sup(\Lambda(ax))^*(t),$$

where the supremum runs over the set of functions $a \in \mathbf{L}_\infty$, $\|a\|_{L_\infty} \le 1$ and the multipliers Λ such that $|\lambda_{n,k}| \le 1$ for every $(n,k) \in \Omega$. It follows immediately from (13.1) that

$$G_t(x) \le C \int_0^1 \frac{x^*(s)}{t+s}\, ds.$$

Theorem 13.1. *Let $x \in L_1, 0 < t \le 1$. Then*

$$G_t(x) \ge \frac{1}{4t} \int_0^t x(s)\, ds.$$

Proof. Given $t \in (0,1]$, denote

$$\gamma = \frac{1}{2t} \int_0^t x^*(s)\, ds$$

and

$$A = \{s : s \in [0,1], | x(s) | > \gamma\}.$$

Prove that

$$\int_A | x(s) |\, ds \ge \gamma t. \tag{13.2}$$

If $B \subset [0,1] \backslash A$ and $mes\, B \le t$, then

$$\int_B | x(s) |\, ds \le \gamma t.$$

Therefore

$$\int_0^t x^*\, ds \le \int_A | x(s) |\, ds + \sup_{\substack{B \subset [0,1]\backslash A \\ mes\, B \le t}} \int_B | x(s) |\, ds \le \int_A | x(s) |\, ds + \gamma t.$$

Hence,

$$\int_A | x(s) |\, ds \ge \int_0^t x^*(s)\, ds - \gamma t \ge 2\gamma t - \gamma t = \gamma t.$$

Denote by D the set of dyadic intervals such that

$$\frac{1}{mes\, d} \int_d | x(s) |\, ds > \gamma.$$

We choose $d_1 \in D$ such that

$$mes\, d_1 = \max_{d \in D} mes\, d.$$

If $d_1, d_2, ..., d_n$ are defined, then $d_{n+1} \in D$ is chosen such that $d_i \cap d_{n+1} = \emptyset$ for every $i = 1, 2, ..., n$ and $mes\, d_{n+1}$ is maximal. It is easy to prove that

$$mes\,(A \setminus \cup_i d_i) = 0. \tag{13.3}$$

Given an index $i \in \mathbf{N}$, denote by e_i the dyadic interval such that $e_i \supset d_i$ and $mes\, e_i = 2 mes\, d_i$. Since $e_i \notin D$, then

$$\frac{1}{mes\, e_i} \int_{e_i} | x(s) | \, ds \leq \gamma.$$

Using this inequality and (13.2), (13.3), we get

$$\begin{aligned} \textstyle\sum_i mes\, d_i = \frac{1}{2} \sum_i mes\, e_i \geq \frac{1}{2\gamma} \sum_i \int_{e_i} | x(s) | \, ds \geq \\ \geq \textstyle\frac{1}{2\gamma} \sum_i \int_{d_i} | x(s) | \, ds \geq \frac{1}{2\gamma} \int_A | x(s) | \, ds \geq \frac{\gamma t}{2\gamma} = \frac{t}{2}. \end{aligned} \tag{13.4}$$

Denote by d_i^+ and d_i^- the left and the right halves of $d_i, i \in \mathbf{N}$. There exists a sequence of integers $\{m_i\}$ such that $d_i = \Delta_{m_i}, i \in \mathbf{N}$. We choose a function $a(t)$ and a sequence $\{\lambda_n\}$ such that

$$a(t)x(t) = \begin{cases} |x(t)|, & t \in \ d_i^+; \\ -|x(t)|, & t \in \ d_i^-; \\ 0, & t \notin \ \cup_i d_i; \end{cases}$$

$$\lambda_n = \begin{cases} 1, & n = m_i; \\ 0, & n \notin \cup_i m_i. \end{cases}$$

Then

$$\Lambda(ax)(t) = \sum_i \frac{1}{mes d_i} \int_{d_i} | x(s) | \, ds \chi_{m_i}$$

and

$$| \Lambda(ax)(t) | \geq \gamma$$

for $t \in \cup_i d_i$. By (13.4),

$$(\Lambda(ax))^*(\frac{t}{2}) \geq \gamma = \frac{1}{2t} \int_0^t x^*(s) \, ds. \tag{13.5}$$

Hence, for every $t \in (0, 1]$,

$$(\Lambda(ax))^*(t) \geq \frac{1}{4t} \int_0^{2t} x^*(s) \, ds \geq \frac{1}{4t} \int_0^t x^*(s) \, ds. \quad \square$$

The function a and the multiplier Λ constructed in Theorem 13.1 depend on $t \in (0, 1]$ in general. Under some assumption on x it is possible to choose a and Λ such that estimate (13.5) is satisfied for every $t \in (0, 1]$ (with an other constant).

Theorem 13.2. *Let x be a non-negative and non-increasing function on $[0,1]$, $x \in L_1$. There exists a multiplier Λ such that $\lambda_{n,k} = 0$ or 1 for every $(n,k) \in \Omega$ and*

$$\Lambda(ax)(t) \geq \frac{1}{240t} \int_0^t x(s)\,ds,$$

where

$$a(t) = \begin{cases} 0, & t \in \cup_{j=0}^{\infty}(2^{-2j-1}, 2^{-2j}) \\ 1, & t \notin \cup_{j=0}^{\infty}(2^{-2j-1}, 2^{-2j}) \end{cases}. \tag{13.6}$$

Proof. We construct a sequence $\{n_i\}$ by induction. Put $n_1 = 0$ and suppose that $n_1, n_2, \ldots, n_{i-1}$ are defined. Let n_i be an even number, $n_i > n_{i-1}$,

$$2^{n_i} \int_0^{2^{-n_i}} x(s)\,ds \geq 5 \cdot 2^{n_{i-1}} \int_0^{2^{-n_{i-1}}} x(s)\,ds, \tag{13.7}$$

and n_i is minimal. The minimality of n_i implies that

$$2^{n_i} \int_0^{2^{-n_i}} x(s)\,ds < 20 \cdot 2^{n_{i-1}} \int_0^{2^{-n_{i-1}}} x(s)\,ds. \tag{13.8}$$

The sequence $\{n_i\}$ is infinite iff $x \notin L_\infty$.

Put

$$\lambda_{n,k} = \begin{cases} 1, & n = n_i, \quad k = 1; \\ 0, & \text{in other cases.} \end{cases}$$

Then

$$\Lambda(ax)(t) = \sum_i 2^{n_i} \left(\int_0^{2^{-n_i}} a(s)x(s)\chi^1_{n_i}(s)\,ds \right) \chi^1_{n_i}(t).$$

Since

$$\int_0^{2^{-n_i}} a(s)x(s)\chi^1_{n_i}(s)\,ds = \int_0^{2^{-n_i-1}} a(s)x(s)\,ds = \sum_{j=1}^{\infty} \int_{2^{-n_i-2j}}^{2^{-n_i-2j+1}} x(s)\,ds \geq$$

$$\geq \sum_{j=1}^{\infty} \frac{1}{3} \left(\int_{2^{n_i-2j}}^{2^{-n_i-2j+1}} + \int_{2^{n_i-2j+1}}^{3\cdot 2^{-n_i-2j}} + \int_{3\cdot 2^{n_i-2j}}^{2^{-n_i-2j+2}} \right) x(s)\,ds =$$

$$= \frac{1}{3} \int_0^{2^{-n_i}} x(s)\,ds,$$

then, for $t \in (2^{-n_{i+1}}, 2^{-n_i})$,

$$\Lambda(ax)(t) \geq 2^{n_i} \int_0^{2^{-n_i}} x(s)\,ds \left(\frac{1}{3} - \sum_{k=1}^{\infty} (\tfrac{1}{5})^k \right) = \frac{1}{12} 2^{n_i} \int_0^{2^{-n_i}} x(s)\,ds.$$

By (13.8), for $t \in (2^{-n_{i+1}}, 2^{-n_i})$

$$\Lambda(ax)(t) \geq \frac{1}{240} 2^{n_{i+1}} \int_0^{2^{-n_{i+1}}} x(s)\,ds \geq \frac{1}{240t} \int_0^t x(s)\,ds.$$

So, the inequality

$$\Lambda(ax)(t) \geq \frac{1}{240t}\int_0^t x(s)\,ds$$

holds for every $t \in (0,1]$. □

The comparision of Theorems 13.1 and 13.2 leads to the following problem. Can $G_t(x)$ be estimated below via $\int_t^1 x^*(s)\frac{ds}{s}$? The answer to this question is negative. To construct a counterexample we must prove two auxiliary statements. Let r be an integer, $N = 2^r$, the numbers k_i are chosen such that $(N+i, k_i) \in O(N, (i-1)2^{N-r}+1)$ for every $i = 1, 2, ..., N$. Recall that the last condition means that

$$\Delta_{N+1}^{k_i} \subset \Delta_N^{(i-1)2^{N-r}+1}.$$

Put

$$x(t) = \sum_{i=1}^N \kappa_{\Delta_{N+1}^{k_i}}(t)$$

and denote by P the set $(n,k) \in \Omega$ such that

$$(N+i, k_i) \in O(n,k) \tag{13.9}$$

for some $i \in \{1, 2, ..., N\}$.

Lemma 13.3. *Let y be a measurable function on $[0,1]$, $|y(t)| \leq x(t)$ for $t \in [0,1]$. Let Λ be a multiplier, $|\lambda_{n,k}| \leq 1$ for $(n,k) \in \Omega$, $\lambda_{n,k} = 0$ for $(n,k) \notin P$. Then*

$$\|\Lambda y\|_{L_\infty} \leq 3.$$

Proof. Denote by $c_{n,k}$, $(n,k) \in \Omega$ the Fourier-Haar coefficients of y. Then

$$\Lambda y(t) = \sum_{(n,k)\in P} \lambda_{n,k} c_{n,k} \chi_n^k(t).$$

If $n \leq N+i$ and $(n,k) \in O(N, (i-1)2^{N-r}+1)$ for some $i \in \{1, 2, ..., N\}$, then

$$| c_{n,k} | \leq 2^{-N-i+n}.$$

Therefore

$$\Big| \sum_{(n,k)\in P_{n>N}} \lambda_{n,k} c_{n,k} \chi_n^k(t) \Big| \leq \max_{1\leq i\leq N} \sum_{n=N+1}^{N+i} 2^{-N-i+n} = 1. \tag{13.10}$$

For each $t \in [0,1]$, we have

$$\Big| \sum_{(n,k)\in P_{n\leq N}} \lambda_{n,k} c_{n,k} \chi_n^k(t) \Big| \leq \sum_{n=1}^N 2^n \int_0^{2^{-n}} x(s)\,ds + 2\int_0^1 x(s)\,ds.$$

If $r \le n \le N$, then

$$\begin{aligned}(0,2^{-n}) &\supset \Delta^{k_1}_{N+1},\\ (0,2^{-n}) &\not\supset \Delta^{k_i}_{N+i}, \quad i = 2,3,\dots,N.\end{aligned}$$

Consequently,

$$\sum_{n=r}^{N} 2^n \int_0^{2^{-n}} x(s)\,ds \le \sum_{n=r}^{N} 2^{n-N-1} < 2^{r-N} < 1. \tag{13.11}$$

If $1 \le n \le r$, then

$$\int_{\Delta^1_{n+1}} x(s)\,ds \ge \int_{\Delta^2_{n+1}} x(s)\,ds.$$

Hence,

$$2^n \int_{\Delta^1_n} x(s)\,ds = 2^n(\int_{\Delta^1_{n+1}} x(s)\,ds + \int_{\Delta^2_{n+1}} x(s)\,ds) \le 2^{n+1} \int_{\Delta^1_{n+1}} x(s)\,ds$$

and

$$\sum_{n=1}^{r} 2^n \int_0^{2^{-n}} x(s)\,ds \le r2^{-r} \int_0^{2^{-r}} x(s)\,ds \le r2^{-r}2^{-N} < 1. \tag{13.12}$$

Taking into account 13.10-12, we get

$$| \Lambda y(t) | \le 1 + 1 + 1 = 3. \quad \square$$

Consider the sequence of functions

$$g_i(t) = \begin{cases} -i - \log_2(t - 2^{-i-1}), & 2^{-i-1} < t \le 2^{-i}, \\ 0, & t \notin (2^{-i-1}, 2^{-i}], \end{cases}$$

($i \in \mathbf{N}$) and the triangular matrix

$$b_{Ni} = \frac{d_N}{N + \log_2 N - 1}, \quad 1 \le i \le N, \tag{13.13}$$

where the numbers d_N are chosen such that, for every $N = 1,2,\dots$, the equalities

$$\sum_{i=1}^{N} b_{Ni} = 1$$

are satisfied. Since

$$\sum_{i=1}^{N} \frac{1}{N + \log_2 N - i} \ge \int_{\log_2 N}^{N} \frac{dt}{t} = \ln \frac{N}{\log_2 N},$$

then

$$\lim_{N\to\infty} d_N = 0.$$

Put

$$z_N(t) = \sum_{i=1}^{N} b_{Ni} g_{N+i}(t). \tag{13.14}$$

Lemma 13.4. *For every integer N, we have*

$$z_N(2^{-2N}) \leq d_N.$$

*Proof.*If $1 \leq i \leq N$, then

$$g^*_{N+i}(\frac{2^{-2N}}{N}) = -N - i - \log_2 \frac{2^{-2N}}{N} = -N - i + \log_2 N.$$

The functions $g_{N+i}(1 \leq i \leq N)$ are disjoint. Therefore

$$z^*_N(2^{-2N}) \leq \max_{1\leq i\leq N} b_{Ni} g^*_{N+i}(\tfrac{2^{-2N}}{N}) \leq$$

$$\leq \max_{1\leq i\leq N} \tfrac{d_N}{N+\log_2 N - i}(N + \log_2 N - i) = d_N.$$

□

Let us construct the sequence of functions

$$x_N(t) = \sum_{i=i}^{N} b_{Ni} \kappa_{\Delta^{k_i}_{N+i}}(t),$$

where b_{Ni} are defined by (13.13) and $\{k_i\}$ satisfies condition (13.9).

Theorem 13.5. *For every $N = 3, 4, ..,$*

$$\int_{2^{-2N}}^{1} \frac{x^*_N(t)}{t}\, dt \geq \frac{1}{2} \ln 2,$$

$$\lim_{N\to\infty} G_{2^{-2N}}(x_N) = 0. \tag{13.15}$$

Proof. Since $x^*_N(2^{-N-i}) \geq b_{Ni}$ for each $1 \leq i \leq N$ and

$$\sum_{i=1}^{N-1} b_{Ni} = 1 - b_{NN} \geq \frac{1}{2},$$

then

$$\int_{2^{-2N}}^{1} \frac{x^*_N(t)}{t}\, dt \geq \sum_{i=1}^{N-1} x_N(2^{-N-i}) \int_{2^{-N-i-1}}^{2^{-N-i}} \frac{dt}{t} \geq \sum_{i=1}^{N-1} b_{Ni} \ln 2 \geq \frac{\ln 2}{2}.$$

To prove (13.15) we fix an integer N. Let $| y(t) |\leq x_N(t)$ for every $t \in [0,1]$, and $\{c_{n,k}\}$ are the Fourier-Haar coefficients of y, $| \lambda_{n,k} |\leq 1$ for every $(n,k) \in \Omega$. Then

$$\Lambda y(t) = \sum_{(n,k)\in\Omega} \lambda_{n,k} c_{n,k} \chi_n^k(t).$$

We construct the set P in accordance with definition (13.9). By Lemma 13.1,

$$| \sum_{(n,k)\in P} \lambda_{n,k} c_{n,k} \chi_n^k(t) |\leq 3d_N. \tag{13.16}$$

Given $i \in \{1,2,...,N\}$, denote

$$V_i(t) = \sum_{(n,k)\in O(N+i,k_i)} \lambda_{n,k} c_{n,k} \chi_n^k(t).$$

Then

$$\sum_{(n,k)\in\Omega\setminus P} \lambda_{n,k} c_{n,k} \chi_n^k(t) = \sum_{i=1}^{N} V_i(t).$$

Applying estimate (13.1) for $\kappa_{\Delta_{N+i}^{k_i}}(t)$, we get

$$V_i^*(t) \leq Cb_{Ni}(-i-N+1-\ln t), \quad 0 < t \leq 2^{-i-N}, \quad 1 \leq i \leq N.$$

Using definition (13.14), we have

$$mes\,\{t :| \frac{1}{2C}\sum_{i=1}^{N} V_i(t) |> \tau\} \leq mes\,\{t : z_N(t) > \tau\}.$$

The obtained inequality and Lemma 13.2 show that

$$(\sum_{i=1}^{N} V_i)^*(2^{-2N}) \leq 2Cd_N. \tag{13.17}$$

Uniting (13.16) and (13.17), we get

$$(\Lambda y)^*(2^{-2N}) \leq (3+2C)d_N.$$

Hence,

$$G_{2^{-2N}}(x_N) \leq (3+2C)d_N.$$

The equality

$$\lim_{N\to\infty} d_N = 0$$

leads to (13.15). □

It was mentioned at the begining of this section that (13.1) implies the inequality

$$G_t(x) \le C\int_0^1 \frac{x^*(s)}{t+s}\,ds \le C\Big(\frac{1}{t}\int_0^t x^*(s)\,ds + \int_t^1 \frac{x^*(s)}{s}\,ds\Big).$$

By Theorem 13.1,

$$G_t(x) \ge \frac{1}{4t}\int_0^t x^*(s)\,ds.$$

Theorem 13.5 shows that it is impossible to estimate $G_t(x)$ below via $\int_t^1 x^*(s)\frac{ds}{s}$. However this estimate takes place under some additional assumption on x.

Theorem 13.6. *Let x be a non-negative and non-increasing functions on $[0,1]$. Then*

$$G_t(x) \ge \frac{1}{24}\int_t^1 x^*(s)\frac{ds}{s}. \tag{13.18}$$

*Proof.*Given an integer m, we put

$$a(s) = \begin{cases} 1 & \text{if } s \in (2^{-m+2j-2}, 2^{-m+2j-1}),\ 0 \le j \le [\frac{m+1}{2}], \\ 0 & \text{for other } s \in [0,1]. \end{cases}$$

$$\lambda_{n,k} = \begin{cases} 1, & 0 \le n < m,\ m-n \ \text{ is even},\ k=1, \\ \varepsilon, & n = m, \quad k = 1, \\ 0, & \text{for other cases}, \end{cases}$$

where $\varepsilon = 0$ or 1 will be chosen later. If $0 \le j \le [\frac{m+1}{2}]$, then

$$\begin{aligned} c_{m-2j,1}(ax) &= 2^{m-2j}\int_0^{2^{2j-m-1}} a(s)x(s)\,ds \ge 2^{m-2j}\int_{2^{2j-m-2}}^{2^{2j-m-1}} x(s)\,ds \ge \\ &\ge 2^{m-2j}2^{2j-m-2}x(2^{2j-m-1}) = \tfrac{1}{4}x(2^{2j-m-1}). \end{aligned}$$

For every $a, b \in R^1$,

$$\max(|\,a+b\,|, |\,a\,|) \ge \frac{1}{3}(|\,a\,| + |\,b\,|).$$

Hence,

$$\max_{\varepsilon=0,1}\min_{0<t<2^{-m}} |\,\Lambda(ax)(t)\,| = \max_{\varepsilon=0,1} |\,-\varepsilon c_{m,1} + \textstyle\sum_{j=1}^{[\frac{m+1}{2}]} c_{m-2j,1}\,| \ge$$

$$\ge \tfrac{1}{3}\textstyle\sum_{j=0}^{[\frac{m+1}{2}]} c_{m-2j,1} \ge \tfrac{1}{12}\sum_{j=0}^{[\frac{m+1}{2}]} x(2^{2j-m-1}) \ge \tfrac{1}{24}\sum_{j=0}^{m} x(2^{-j-m-1}) \ge$$

$$\ge \tfrac{1}{24}\textstyle\sum_{j=0}^{m}\int_{2^{j-m-1}}^{2^{j-m}} \frac{x(s)}{s}\,ds = \tfrac{1}{24}\int_{2^{-m-1}}^{1} \frac{x(s)}{s}\,ds.$$

If $t \in (0,1)$ we can choose an integer m such that $2^{-m-1} < t \leq 2^{-m}$. Then

$$G_t(x) \geq G_{2^{-m}}(x) \geq \frac{1}{24}\int_{2^{-m-1}}^{1} \frac{x(s)}{s}\,ds \geq \frac{1}{24}\int_t^1 \frac{x(s)}{s}\,ds. \quad \square$$

Comparing Theorems 13.1 and 13.6, we get

Corollary 13.7. *Let x be a non-negative and non-increasing function on [0,1]. Then*

$$G_t(x) \geq \frac{1}{28}\int_0^1 \frac{x(s)}{t+s}\,ds.$$

This means that

$$G_t(x) \approx \int_0^1 \frac{x(s)}{t+s}\,ds$$

on the set of non-negative and non-increasing functions. The equivalence constants are absolute. We have proved in Chapter 6 that the H.s. forms an unconditional basis in a separable r.i. space E iff

$$0 < \alpha_E \leq \beta_E < 1.$$

The sufficiensy follows immediately from the interpolation theorem. The necessity proof was complicated. Using the obtained results it is possible to present a more simple proof.

Proof of the necessity of Theorem 6.6. Denote by K the unconditional constant of the H.s. in E. Let $y(t)$ be a non-negative and non-decreasing function on [0,1]. By Theorem 13.2, there exist a sequence $\lambda_{n,k} = 0$ or 1 for $(n,k) \in \Omega$ and a function $a(t)$ such that $a(t) = 0$ or 1 and

$$\Lambda(ay)(t) \geq \frac{1}{240}\frac{1}{t}\int_0^t y(s)\,ds.$$

Hence,

$$\|Hy\|_E \leq 240\|\Lambda(ay)\|_E \leq 240K\|y\|_E, \tag{13.19}$$

where

$$Hx(t) = \frac{1}{t}\int_0^t x(s)\,ds.$$

Clearly, the norm of H is attained on the set of non-negative and non-increasing functions. This follows from (1.c.2). Therefore (13.19) implies that H is bounded in E and

$$\|H\|_E \leq 240K.$$

Applying Theorem 1.d.2, we get $\alpha_E > 0$. By duality arguments, $\beta_E < 1$. $\square$

Chapter 14

Estimates of Multipliers in L_1

Here we use the notations of Chapter 12, 13. If $|\lambda_{n,k}| \leq 1$ for $(n,k) \in \Omega$, then the corresponding multiplier Λ is of week type $(1,1)$ and its norm is less than 2. This statement was proved in Chapter 5. This chapter is devoted to the unimprovability of the mentioned result.

Denote by $L_{1,\infty}$ the set of measureable on $[0,1]$ functions such that

$$\|x\|_{L_{1,\infty}} = \sup_{0<t\leq 1} x^*(t)\,t < \infty.$$

It is well know that $L_{1,\infty}$ is a quasi-normed space and the triangle inequality in $L_{1,\infty}$ is satisfied for the constant 2. The topology of $L_{1,\infty}$ is not normed.

Let us consider the functional

$$F(x) = \inf \|\Lambda(ax)\|_{L_1}$$

on L_1, where the infinum is taken over the set of functions $a \in L_\infty$, $|a(t)| \geq 1$ for every $t \in [0,1]$ and multipliers Λ such that $|\lambda_{n,k}| = 1$ for every $(n,k) \in \Omega$. Theorem 5.1 implies the inequality

$$\frac{1}{2}\|x\|_{L_{1,\infty}} \leq F(x).$$

Theorem 14.1. *Let $x(t)$ be a non-negative and non-increasing function, $x \in L_1$. Then*

$$F(x) \leq 4\|x\|_{L_{1,\infty}}. \tag{14.1}$$

Proof. Let $\|x\|_{L_{1,\infty}} \leq 1$ and $x(1) > 0$. For $t \in (0,1]$, we have $x(t) \leq t^{-1}$. In fact, $x(2^{-k}) \leq 2^k$ for $k = 0, 1, \ldots$ Since

$$\sum_{k=0}^{\infty} 2^{-k} x(2^{-k}) \leq 2\int_0^1 x(t)\,dt < \infty,$$

then

$$\lim_{k\to} 2^{-k}x(2^{-k}) = 0.$$

We construct the sequence $(\alpha_0, \alpha_1, ...)$ by induction. Put $\alpha_0 = 1$. If $\alpha_0, \alpha_1, ..., \alpha_{n-1}$ are defined, then we put

$$\alpha_n = \max\left(\max_{k\geq n+1} 2^{-k}x(2^{-k}), \frac{\alpha_{n-1}}{2}\right).$$

It is evident that $\alpha_n \downarrow 0$ and $\alpha_n \geq 2^{-n-1}x(2^{-n-1})$. Let us consider the function

$$y(t) = \sum_{n=0}^{\infty}(-2)^n\chi_n^1(t) = -\kappa_{(\frac{1}{2},1)}(t) + \sum_{n=1}^{\infty}\left((\sum_{j=0}^{n-1}(-2)^j\alpha_n\right)\kappa_{(2^{-n-1},2^{-n})}(t).$$

Since

$$|\sum_{j=0}^{n-1}(-2)^j\alpha_j - (-2)^n\alpha_n| = 2^n\alpha_n + 2^{n-1}\alpha_{n-1} - 2^{n-2}\alpha_{n-2} + ... \geq 2^n\alpha_n,$$

then

$$|y(t)| \geq \kappa_{(\frac{1}{2},1)}(t) + \textstyle\sum_{n=1}^{\infty} 2^n\alpha_n\kappa_{2^{-n-1},2^{-n})}(t) \geq \kappa_{(\frac{1}{2},1)}(t) +$$

$$+\tfrac{1}{2}\textstyle\sum_{n=1}^{\infty} x(2^{-n-1})\kappa_{(2^{-n-1},2^{-n})}(t) \geq \tfrac{1}{2}x(t).$$

Therefore there exists a function $a(t)$ such that $|a(t)| \geq 1$ and $a(t)\,x(t) = 2y(t)$. Put

$$\lambda_{n,k} = \begin{cases} -1, & \text{if } n \text{ is even and } k = 1, \\ 1, & \text{in other cases.} \end{cases}$$

Then

$$\Lambda y(t) = \textstyle\sum_{n=0}^{\infty} 2^n\alpha_n\chi_n^1(t) =$$

$$= -\kappa_{(\frac{1}{2},1)}(t) + \textstyle\sum_{n=1}^{\infty}\left((\sum_{j=0}^{n-1} 2^j\alpha_n) - 2^n\alpha_n\right)\kappa_{(2^{-n-1},2^{-n})}(t).$$

For $t \in (2^{-n-1}, 2^{-n})$, $n = 1, ...,$ we have

$$1 + \Lambda y(t) = 1 + \left(\sum_{j=0}^{n-1} 2^j\alpha_j\right) - 2^n\alpha_n \geq \sum_{j=0}^{n-1} 2^j(\alpha_j - \alpha_n) \geq 0.$$

Hence,

$$\|1+\Lambda y\|_{L_1} = \lim_{N\to\infty}\sum_{n=1}^{N}(1+\sum_{j=0}^{n-1}2^j\alpha_j - 2^n\alpha_n)2^{-n-1} =$$

$$= \tfrac{1}{2}+\tfrac{1}{2}\lim_{N\to\infty}\sum_{n=1}^{N}\left(\sum_{j=0}^{n-1}2^{j-n}\alpha_j - \alpha_n\right) =$$

$$= \tfrac{1}{2}+\tfrac{1}{2}\lim_{N\to\infty}\left(\sum_{j=0}^{N-1}(\sum_{n=j+1}^{N}2^{j-n}\alpha_j) - \sum_{n=1}^{N}\alpha_n\right) \le$$

$$\le \tfrac{1}{2}+\tfrac{1}{2}\lim_{N\to\infty}(\sum_{j=0}^{N-1}\alpha_j - \sum_{n=1}^{N}\alpha_n) = \tfrac{1}{2}+\tfrac{\alpha_0}{2} = 1.$$

Consequently,

$$F(x) \le 2\|\Lambda y\|_{L_1} \le 4. \quad \square$$

The assumption on the monotonisity of x is essential. To prove this statement we must establish two lemmas. We denote by V the set of sequences of real numbers $\{c_k\}_0^\infty$ with the bounded variation and tending to 0. We put

$$\|c\|_V = \sum_{k=1}^{\infty}|c_k - c_{k-1}|.$$

Lemma 14.2. *Let $c \in V$. Then*

$$\frac{1}{3}\|c\|_V \le \|\sum_{n=0}^{\infty}c_n 2^n \chi_n^1\|_{L_1} \le 4\|c\|_V. \tag{14.2}$$

Proof. We have

$$\sum_{n=0}^{\infty}c_n 2^n\chi_n^1(t) = -c_0\kappa_{(\frac{1}{2},1)}(t) + \sum_{n=1}^{\infty}(\sum_{k=0}^{n-1}c_k 2^k - c_n 2^n)\kappa_{(2^{-n-1},2^{-n})}(t). \tag{14.3}$$

At first we suppose that $c_0 \ge c_1 \ge \ldots \ge 0$. Then for each $t \in [0,1]$

$$c_0 + \sum_{n=0}^{\infty}c_n 2^n\chi_n^1(t) \ge \sum_{n=1}^{\infty}\sum_{k=0}^{n-1}(c_k - c_n)2^k)\kappa_{(2^{-n-1},2^{-n})}(t) \ge 0.$$

The series

$$\sum_{n=0}^{\infty}c_n 2^n\chi_n^1(t) \tag{14.4}$$

converges in L_1. Indeed,

$$\|\sum_{n=i}^{j}c_n 2^n\chi_n^1\|_{L_1} \le \|c_i 2^i\chi_i^1\|_{L_1} + \|c_i 2^i\kappa_{(0,2^{-i})} + \sum_{n=i}^{j}c_n 2^n\chi_n^1\|_{L_1} =$$

$$= c_i + \int_0^1\left(c_i 2^i\kappa_{(0,2^{-i})}(t) + \sum_{n=i}^{j}c_n 2^n\chi_n^1(t)\right)dt = 2c_i$$

for every $1 \le i < j$. Therefore the assumption

$$\lim_{i\to\infty} c_i = 0$$

implies the convergence of series (14.4) in L_1. So

$$\|\sum_{n=0}^{\infty} c_n 2^n \chi_n^1\|_{L_1} \le 2c_0.$$

If $c \in V$, then it may be represented as a difference of some two monotone sequences. Applying the obtained inequality, we get the right part of (14.2).

Let m be an integer, and

$$\|\sum_{n=0}^{m} c_n 2^n \chi_n^1\| \le 1.$$

By (14.3)

$$\frac{|c_0|}{2} + \sum_{n=1}^{m} 2^{-n-1} |\sum_{k=0}^{n-1} c_k 2^k - c_n 2^n| = \frac{|c_0|}{2} + \frac{1}{2}\sum_{n=1}^{m} |\sum_{k=0}^{n-1} c_k 2^{k-n} - c_n| \le 1.$$

Denote

$$\begin{array}{ll} c_0 & = d_0 \\ c_n & - \sum_{k=0}^{n-1} 2^{k-n} c_k = d_n, \quad 1 \le n \le m. \end{array}$$

Then

$$c_0 = d_0$$

$$c_n = d_n + \tfrac{1}{2} \textstyle\sum_{k=0}^{n-1} d_k, \tag{14.5}$$

$$\textstyle\sum_{n=0}^{m} |d_n| \le 2.$$

Hence,

$$\max_{\|\sum_{n=0}^{m} c_n 2^n \chi_n^1\|_{L_1} \le 1} \|c\|_V \le 2 \max_{\sum_{n=0}^{m} |d_n| \le 1} \|\{d_n + \frac{1}{2}\sum_{k=0}^{n-1} d_k\}\|_V. \tag{14.6}$$

The functional

$$\|\{d_n + \frac{1}{2}\sum_{k=0}^{n-1} d_k\}\|_V$$

is convex on the set

$$\{d : \sum_{n=0}^{m} |d_k| \le 1\}. \tag{14.7}$$

It is well known that the extremal points of set (14.7) are $\{0, 0, ..., \pm 1, 0, ..., 0\}$. By (14.5), (14.6)

$$\|c\|_V \le 2 \max_{0\le n\le m} \|0, 0, ..., 0, \overset{n}{1}, \tfrac{1}{2}, \tfrac{1}{2}, ..., \tfrac{1}{2}\|_V = 2\tfrac{3}{2} = 3.$$

This proves the left part of (14.2) for a finite c. Passing to the limit, when $m \to \infty$, we obtain (14.2) for an arbitrary $c \in V$. □

Lemma 14.3. *Let $\{e_i\}$ be a finite or countable sequence of disjoint subsets of $[0,1]$. If $x(t)$ is a constant on e_i and*

$$\int_{e_i} y(t)\,dt = 0$$

for every $i = 1, 2, ...,$ then

$$\|x + y\|_{L_1} \ge \frac{1}{2} \max(\|x\|_{L_1}, \|y\|_{L_1}).$$

Proof. Clearly, it is sufficient to prove this Lemma for the case $c_i = [0,1], e_2 = e_3 = ... = \emptyset$. Let $x(t) = c > 0$. If $\|y\|_{L_1} \le 2c$, then

$$\|x + y\|_{L_1} \ge \int_0^1 (c + y(t))\,dt = c \ge \frac{1}{2} \max(\|x\|_{L_1}, \|y\|_{L_1}).$$

If $\|y\|_{L_1} \ge 2c$, then

$$\|x+y\|_{L_1} \ge \|y\|_{L_1} - c \ge \|y\|_{L_1} - \frac{1}{2}\|y\|_{L_1} = \frac{1}{2}\|y\|_{L_1} = \frac{1}{2}\max(\|x\|_{L_1}, \|y\|_{L_1}).$$

□

It is evident that the constant $\frac{1}{2}$ in Lemma 14.3 is exact.

Let us consider the sequence of functions

$$x_N(t) = \sum_{k=1}^{N} 2^{2k} \kappa_{(2^{-2k}, 2^{-2k+1})}(t), \quad N = 1, 2, ...$$

Since

$$0 \le x_N(t) \le \frac{2}{t}$$

for $t \in (0,1]$, then $\|x_N\|_{L_{1,\infty}} \le 2$.

Theorem 14.4. *For any integer N*

$$F(x_N) \ge \frac{N-1}{24}.$$

Proof. We fix a measurable bounded function $a(t)$, $|a(t)| \geq 1$ and a sequence $\lambda_{n,k} = \pm 1, (n,k) \in \Omega$ and prove that

$$\|\Lambda(a x_N)\|_{L_1} \geq \frac{N-1}{24},$$

where Λ is the corresponding multiplier. Given $k \in \{1, 2, ..., N\}$, denote

$$b_k(t) = a(t)\kappa_{(2^{-2k}, 2^{-2k+1})}(t),$$

$$\mu_k = 2^{2k} \int_{2^{-2k}}^{2^{-2k+1}} a(t)\, dt,$$

$$f(t) = \sum_{k=1}^{N} (a(t) - \mu_k)\kappa_{(2^{-2k}, 2^{-2k+1})}(t) = \sum_{k=1}^{N} f_k(t),$$

$$g(t) = \sum_{k=1}^{N} \mu_k \kappa_{(2^{-2k}, 2^{-2k+1})}(t).$$

Then

$$a(t)\, x_N(t) = f(t) + g(t).$$

Since

$$supp\, f_k \subset (2^{-2k}, 2^{-2k+1})$$

and

$$\int_{2^{-2k}}^{2^{-2k+1}} f_k(t)\, dt = 0,$$

then

$$supp\, \Lambda f_k \subset (2^{-2k}, 2^{-2k+1}), \quad k = 1, 2, ..., N$$

Denote

$$Q = \{k : 1 \leq k \leq N, \ | \mu_k | \leq 1\}.$$

If $k \in Q$, then by Theorem 5.1

$$\|\Lambda f_k\|_{L_1} \geq \frac{1}{2}\|f_k\|_{L_{1,\infty}} \geq (1 - | \mu_k |)2^{2k}2^{-2k-1} = \frac{1}{2}(1 - | \mu_k |).$$

The functions

$$\Lambda f_k(t), \quad 1 \leq k \leq N$$

are disjoint. Therefore

$$\|\Lambda f\|_{L_1} \geq \frac{1}{2} \sum_{k \in Q} (1 - | \mu_k |). \tag{14.8}$$

Let us check the Fourier-Haar coefficients $c_{ij} = c_{ij}(g)$ of the function g. Clearly, $c_{00} = c_{01} = \sum_{n=1}^{N} \mu_n$,

$$c_{2k-1,1} = 2^{2k-1}(-\mu_k + \sum_{n=k+1}^{N} \mu_n), \quad 1 \leq k \leq N,$$

$$c_{2k,1} = 2^{2k} \sum_{n=k+1}^{N} \mu_n, \quad 1 \leq k \leq N-1$$

and $c_{ij} = 0$ if $j > 1$ or $j = 1$ and $i \geq 2N$. Hence,

$$\Lambda g(t) = \lambda_{00} c_{00} \chi_0^0(t) + \sum_{i=0}^{2N-1} \lambda_{i,1} c_{i,1} \chi_i^1(t).$$

Applying Lemmas 14.3 and 14.2, we get

$$\|\Lambda g\|_{L_1} \geq \tfrac{1}{2} \| \textstyle\sum_{i=0}^{2N-1} \lambda_{i,1} c_{i,1} \chi_i^1 \|_{L_1} \geq \tfrac{1}{6} \textstyle\sum_{i=1}^{2N-1} | 2^{-i} \lambda_{i,1} c_{i,1} -$$

$$-2^{1-i} \lambda_{i-1,1} c_{i-1,1} | \geq \tfrac{1}{6} \textstyle\sum_{k=1}^{N-1} (| 2^{2-2k} \lambda_{2k-2,1} c_{2k-2,1} -$$

$$-2^{1-2k} \lambda_{2k-1,1} c_{2k-1,1} | + | 2^{1-2k} \lambda_{2k-1,1} c_{2k-1,1} -$$

$$-2^{-2k} \lambda_{2k,1} c_{2k,1} |) = \tfrac{1}{6} \textstyle\sum_{k=1}^{N-1} (| \lambda_{2k-2,1} \sum_{n=k}^{N} \mu_n -$$

$$-\lambda_{2k-1,1} (-\mu_k + \textstyle\sum_{n=k+1}^{N} \mu_n) | + | \lambda_{2k-1,1} (-\mu_k +$$

$$+ \textstyle\sum_{n=k+1}^{N} \mu_n) - \lambda_{2k,1} \sum_{n=k+1}^{N} \mu_n |) \geq$$

$$\geq \tfrac{1}{6} \textstyle\sum_{k=1}^{N-1} \Big(\min(2 | \mu_k |, 2 \sum_{n=k+1}^{N} | \mu_n | +$$

$$+ \min(| \mu_k |, | \mu_k - 2 \textstyle\sum_{n=k+1}^{N} \mu_n |) \Big).$$

Since $2 \min(| a |, | b |) + \min(| a |, | a - 2b |) \geq | a |$ for every $a, b \in R^1$, then

$$\|\Lambda g\|_{L_1} \geq \frac{1}{6} \sum_{k=1}^{N-1} | \mu_k | . \tag{14.9}$$

The function $\Lambda g(t)$ is a constant on each interval $(2^{-i}, 2^{-i+1})$ and

$$\int_{2^{-i}}^{2^{-i+1}} \Lambda g(t)\, dt = 0.$$

By Lemma 14.3

$$\|\Lambda(a x_N \|_{L_1} \geq \frac{1}{2} \max(\|\Lambda f\|_{L_1}, \|\Lambda g\|_{L_1}) \geq \frac{1}{4}(\|\Lambda f\|_{L_1} + \|\Lambda g\|_{L_1}).$$

Using estimates (14.8) and (14.9), we have

$$\|\Lambda(a x_N)\|_{L_1} \geq \frac{1}{4} \left(\frac{1}{2} \sum_{k \in Q} (1 - | \mu_k |) + \frac{1}{6} \sum_{k=1}^{N-1} | \mu_k | \right) \geq \frac{1}{24} \sum_{k=1}^{N-1} 1 = \frac{N-1}{24}.$$

□

Theorem 14.4 shows that to estimate $F(x)$ above through $\|x\|_{L_{1,\infty}}$ is impossible.

Chapter 15

Subsequences of the Haar system

If the H.s. is an unconditional basis of an r.i. space E, then the spaces spanned by subsequences of the H.s. are complemented in E. These spaces can be characterized in the following form.

Theorem 15.1. *Let E be an r.i. space, $0 < \alpha_E \leq \beta_E < 1$. Let $\{x_n\}_{n=1}^{\infty}$ be a subsequence of the H.s. and $B := \cap_{n=1}^{\infty} \cup_{k=n}^{\infty} supp\, x_k$. If $mes(B) = 0$, then there exists a sequence of pairwise disjoint characteristic functions, whose closed linear span contains $[x_n]_{n=1}^{\infty}$. If $mes(B) > 0$, then $[x_n]_{n=1}^{\infty}$ is isomorphic to E.*

Proof. Let us investigate first the case $mes(B) = 0$. Put

$$J_n = \{k = 1, 2, \dots : \; supp\, x_k \subset supp\, x_n, \; k \neq n\},$$

$$B_n = \{t : x_n(t) > 0\} \setminus \cup_{k \in J_n} supp\, x_k,$$

$$C_n = \{t : x_n(t) < 0\} \setminus \cup_{k \in J_n} supp\, x_k, \quad n \in \mathbf{N}.$$

It is clear that

$$\cup_{n=1}^{\infty} (B_n \cup C_n) = \cup_{k=1}^{\infty} supp\, x_k \setminus B,$$

$$B_n \cap B_k = B_n \cap C_k = C_n \cap C_k = \emptyset, \quad n \neq k; \quad n, k \in \mathbf{N}.$$

Since $mes(B) = 0$ it follows that

$$[\kappa_{B_n}, \kappa_{C_n}]_{n=1}^{\infty} \supset [x_n]_{n=1}^{\infty}.$$

Let $mes(B) > 0$. Denote $X := [x_n]_{n=1}^{\infty}$. To prove the statement it suffices to find a dyadic tree $\{A_n^k\}_{(n,k) \in \Omega}$ and a sequence $\{y_m\}_{m=2}^{\infty}$, $y \in X$ such that

$$A_0^1 = B, \; m(A_n^k) = 2^{-n} m(B), \; (n, k) \in \Omega;$$

$$\|y_m - H_m\|_E \leq \delta_m, \quad m = 2, 3, \dots,$$

where $\{H_m\}_{m=1}^{\infty}$ is a generalized H.s. associated with $\{A_n^k\}_{(n,k)\in\Omega}$, and $\{\delta_m\}_{m=2}^{\infty}$ is a sequence of scalars for which

$$\sum_{m=2}^{\infty} \frac{\delta_m}{\|H_m\|_E} < \frac{1}{8}.$$

Indeed, the subspace $Z := [H_m]_{m=2}^{\infty}$ is isomorphic to E and complemented in E. By Theorem 1.b.2 the subspace $Y := [y_m]_{m=2}^{\infty}$ has the same properties. Hence, by Theorem 1.d.4 X is isomorphic to E.

To construct A_1^1 we fix $\gamma > 0$ and find an open subset $G \subset B$ such that $mes(G \backslash B) \leq \gamma$. For convenience we assume that the enumeration of the functions in $\{x_n\}_1^{\infty}$ corresponds to their order in the H.s. For every $t \in B$, let $n(t)$ be the smallest integer $n \neq 1$ for which $t \in supp\, x_n \subset G$. The existence of such numbers follows from the definition of B and the openness of G. It is clear that the distinct functions among $x_{n(t)}, t \in B$ have mutually disjoint supports and thus their sum y_2 belongs to X. Furthermore, $\kappa_B \leq | y_2 | \leq \kappa_G$. If γ is small enough, then it is possible to find $A_1^1 \subset B, mes(A_1^1) = 2^{-1} mes(B)$ such that the generalized Haar function $H_2 = H_0^1$ satisfies $\|y_2 - H_2\|_E \leq \delta_2$.

By replacing in the above construction B by A_1^1, we can find the function $y_3 \in X$ and the sets A_2^1, A_2^2 of the dyadic tree with $mes(A_2^1) = mes(A_2^2) = 2^{-2} mes(B)$ such that $\|y_3 - H_3\|_E \leq \delta_3$. The construction of the entire tree and of the sequence $\{y_m\}_{m=2}^{\infty}$ is continued by inductive argument. □

Corollary 15.2. *Every subsequence of the H.s. in L_p, $1 < p < \infty$, spans a subspace, which is isomorphic either to l_p or to L_p.*

Proof. It is necessary to investigate only the case $mes(B) = 0$. The statement follows from the following two facts. Every sequence of pairwise disjoint characteristic functions spans in L_p the subspace which is isomorphic to l_p. Every complemented subspace of l_p is isomorphic to l_p [162, p.54] □

If $p = 1$, Corollary 15.2 is not valid. Moreover, using the results from [8], it is possible to construct the subsequence of the H.s. whose closed linear span in L_1 is not an $\mathcal{L}_1$-space. On the other hand, by combining the argument of Theorem 15.1 with Remark 1.d.5, we obtain that every complemented subspace of L_1 spanned by a subsequence of the H.s. is isomorphic either to l_1 or to L_1. Thus, it is interesting to characterize subsequences of the H.s. whose closed linear spans are complemented in L_1.

Later on we assume for convenience that the functions of the subsequence are enumerated according to their order in the H.s.

Definition 15.3. *A subsequence of the H.s. $\{x_n\}_1^{\infty}$ is said to be without branchings, if for any n each of the sets*

$$\{t : x_n(t) > 0\} \cap (\cap_{i=n+1}^{\infty} supp\, x_i),$$

$$\{t : x_n(t) < 0\} \cap (\cap_{i=n+1}^{\infty} supp\, x_i)$$

is either empty or equal to $supp\, x_l$ *for some* $l > n$.

Theorem 15.4. *If* $\{x_n\}_1^\infty$ *is a subsequence of the H.s. without branchings, then* $[x_n]_1^\infty$ *is a complemented subspace of* L_1.

Before proving this theorem, let us introduce two definitions.

Definition 15.5. *A group of the Haar functions* K *is called a bush, if it is possible to enumerate the elements of* K *in such a way that:*

$$K = \{y_n\}_1^{2^m-1}, \quad m = 2, 3, ...;$$

$$\{t : y_{2^i+j}(t) > 0\} = supp\, y_{2^{i+1}+2j};$$

$$\{t : y_{2^i+j}(t) < 0\} = supp\, y_{2^{i+1}+2j+1};$$

where $0 \le j \le 2^i - 1, \quad 0 \le i \le m-2.$

Let $\{A_i^K\}_{i=1}^{2^m}$ be the atoms of the minimal algebra of subsets of $supp\, y_1$ with respect to which all the functions of K are measurable. It is clear that

$$[\kappa_{supp\, y_1} \cup K] = [\kappa_{A_i^K}]_{i=1}^{2^m}, \tag{15.1}$$

and for every $r \in \{1, 2, ..., 2^m\}$

$$[K] = [\kappa_{A_i^K} - \kappa_{A_r^K},\ i = 1, 2, ..., 2^m;\ i \neq r]. \tag{15.2}$$

The notion of a bush can be naturally generalized on the infinite group of the Haar fucntions. In this case we have that

$$[\kappa_{supp\, y_1} \cup K] = L_1(supp\, y_1),$$

$$[K] = \{f \in L_1(supp\, y_1) : \int_{supp\, y_1} f(t)\, dt = 0\}.$$

Definition 15.6. *A group of the Haar functions* B *is called a branch if the elements of* B *can be enumerated in the following way:*

$$B = \{z_i\}_{i=1}^m; \quad supp\, z_{i+1} \subset supp\, z_i;$$

$$mes(supp\, z_{i+1}) = \frac{1}{2} mes(supp\, z_i),$$

where $1 \le i < m,\ m \in \mathbf{N}.$

Let $\{A_i^B\}_{i=1}^{m+1}$ be the atoms of the minimal algebra of subsets of $supp\, z_1$ with respect to which all functions of B are measurable. For convenience we assume that $mesA_{i+1}^B \leq mesA_i^B,\ i \in \{1,2,...,m\}$. It is easy to show that

$$[\kappa_{supp\, z_1} \cup B] = [\kappa_{A_i^B}]_{i=1}^{m+1}, \tag{15.3}$$

$$[B] = [2^{m-1}\kappa_{A_{m+1}^B} - \kappa_{A_1^B}, 2^{i-1}\kappa_{A_i^B} - \kappa_{A_1^B}, \quad i = 2,...,m]. \tag{15.4}$$

Proof of Theorem 15.4. Let $\{x_k\}_1^\infty$ be a subsequence of the H.s. without branchings. Let us investigate first the subsequences $\{x_k\}_1^\infty$ which contain no infinite bush. For convenience we assume that $\chi_1 \notin \{x_k\}_1^\infty$. The scheme of the proof is the following :

1) for every n, we construct functions $\{y_k^n\}_{k=1}^n$ such that $[y_k^n]_{k=1}^n = [x_k]_{k=1}^n$; these functions allow us to define the projector $P_n : L_1 \to [x_k]_1^n$ with $\|P_n\|_{L_1} \leq 4, n \in \mathbf{N}$.

2) it occurs that for every $x \in L_1$ the sequence $\{P_n x\}_{n=1}^\infty$ is Cauchy one in L_1; hence, we have a projector $P : L_1 \to [x_k]_{k=1}^\infty$ with $\|P\|_{L_1} \leq 4$.

Fix a natural number n. The construction of $\{y_k^n\}_{k=1}^n$ is fulfilled in two steps.

Step 1. Let K be the maximal (with respect to the number of elements) bush containing x_1. If such K exists, then we pass to the consideration of x_l, where $l = \min\{k = 2,...,n : x_k \notin K\}$. For x_l we find the maximal bush L containing x_l. If K does not exist, then we determine the maximal bush containing x_2. In general, after the recurrent step of defining the maximal bushes, we pass to the consideration of the disregarded (i.e. not considered yet, and which is not contained in bushes, defined before) function with the minimal number. If it is possible, then we find the maximal bush containing this function. When this process is finished, the functions, which are not contained in any bush, are divided in the maximal (with respect to the number of elements) branches. The obtained groups of the functions are enumerated according to the order of increasing of the minimal index of the elements of the group. Finally, we have $\{x_k\}_{k=1}^n = \{X_i\}_{i=1}^m$, where X_i is either a bush or a branch. For convenience, instead of the notation $A_j^{X_i}$ from Definitions 15.5 and 15.6, $j \in \{1,2,...,card\, X_i + 1\}$; $i \in \{1,2,...,m\}$ we shall use the notation A_{ij}.

Let us consider X_1. Due to equations (15.2) or (15.4) there exists a group of functions with a linear span equal to $[X_1]$. Furthermore, we can choose the index $r \in \{1,2,..., card\, X_1 + 1\}$ in equation (15.2) (if X_1 is a bush) in such a way that

$$mes(\cup_{\{j=1,2,...,n: supp\, x_j \subset A_{1r}\}} supp\, x_j) \leq \frac{1}{2} mes(A_{1r}). \tag{15.5}$$

Such index r exists due to the following argument. By definition 15.3 the fraction

$$\frac{mes(\cup_{\{j=1,2,...,n: supp\, x_j \subset A_{1r}\}} supp\, x_j)}{mes(A_{jr})}$$

is equal to either 0 or $2^{-k}, k = 0, 1, \ldots$ That is why from the assumption that there does not exist an index r satisfying (15.5), it follows that the bush X_1 is not maximal with respect to the number of elements.

After normalization in L_1 the group of functions from equations (15.2) or (15.4) corresponding to X_1 is denoted by W_1. The construction is continued by inductive argument. Assume that $\cup_{i=1}^{l} X_l, 1 \leq l < m$, has been already considered, and we have constructed W_l. For X_{l+1} two cases are possible.

I. For every $i \in \{1, 2, .., l\}$ and $j \in \{1, 2, \ldots card\, X_i + 1\}$

$$\cup_{x \in X_{l+1}} supp\, x \neq A_{ij}.$$

In this case we define the new group of functions in exactly the same way as for X_1, and W_{l+1} is equal to the union of W_l with this new group.

II. For some $i \in \{1, 2, \ldots, l\}$ and some $j \in \{1, 2, .., card\, X_i + 1\}$

$$\cup_{x \in X_{l+1}} supp\, x = A_{ij}.$$

In this case we find first the function $z \in W_l$ such that $A_{ij} = \{t : z(t) > 0\}$, and afterwards we define

$$W_{l+1} = (W_l \backslash \{z\}) \cup \{(2mes(A_{l+1,p}))^{-1} \kappa_{A_{l+1,p}} + z\kappa_{(0,1)\backslash A_{ij}}\}_{p=1}^{card\, X_{l+1}+1}.$$

After consideration of X_m we obtain the group of functions W_m. The elements of W_m are enumerated according to the following rule. First of all, we enumerate in an arbitrary way the functions $z \in W_m$, for which there exists no function $z' \in W_m$ with $supp\, z' \supset \{t : z(t) > 0\}$. Denote the set of functions enumerated at the first step as W_m^1. Then, we enumerate the functions $z \in W_m \backslash W_m^1$ for which there exists no function $z' \in W_m \backslash W_m^1$ with $supp\, z' \supset \{t : z(t) > 0\}$. The enumeration is continued similarly until the exhaustion of W_m. The obtained sequence is denoted as $\{z_k^n\}_{k=1}^n = W_m$.

Put

$$J_k^n = \{i \in \{1, 2, \ldots, n\} : supp\, z_i^n \subset \{t : z_k^n(t) > 0\}\},$$

$$B_k^n = \{t : z_k^n(t) > 0\} \backslash \cup_{i \in J_k^n} supp\, z_i^n, \quad k = 1, 2, \ldots, n.$$

By the construction, for every $k = 1, 2, \ldots, n$

$$mes(B_k^n) \geq 2^{-1} mes\{t : z_k^n(t) > 0\},$$

and

$$\int_{B_k^n} |z_k^n(t)|\, dt \geq \frac{1}{4}.$$

Furthermore, if $l < k$ and/or $supp\, z_l^n \cap supp\, z_k^n = \emptyset$; $l, k = 1, 2, \ldots, n$, then $supp\, z_k^n \cap B_l^n = \emptyset$. It is obvious that $[z_k^n]_{k=1}^n = [x_k]_{k=1}^n$, $\|z_k^n\|_{L_1} = 1$, $k = 1, \ldots, n$.

Before continuing the proof let us illustrate the construction on the example: $\{x_k\}_1^6 = \{\chi_0^1, \chi_1^1 \cdot \chi_2^1, \chi_2^2, \chi_2^3, \chi_3^5\}$. Then, $X_1 = \{\chi_0^1\}$ is a branch, $X_2 = \{\chi_1^1, \chi_2^1, \chi_2^2\}$ is a bush, $X_3 = \{\chi_2^3, \chi_3^5\}$ is a branch too. The functions $\{z_k^6\}_1^6$ have the following form: for $k = 1, 2, 3, 4$,

$$z_k^6 = 4\kappa_{(\frac{k-1}{8}, \frac{k}{8})} - \kappa_{(\frac{1}{2}, 1)};$$

for $k = 5, 6$,

$$z_k^6 = 8\kappa_{(\frac{1}{2}+\frac{k-5}{16}, \frac{1}{2}+\frac{k-4}{16})} - 4\kappa_{(\frac{5}{8}, \frac{3}{4})}.$$

Now we are going to construct the functions $\{y_k^n\}_{k=1}^n$ with the following properties:

$$\int_{B_k^n} | y_k^n(t) | \, dt \geq \frac{1}{4}, \tag{15.6}$$

and

$$supp\, y_k^n \cap B_l^n = \emptyset, \quad l \notin k; \quad l, k = 1, 2, ..., n; \tag{15.7}$$

where the sets B_k^n have been defined before.

We shall give the description of the construction of y_1^n; the rest of the mentioned above functions can be defined in exactly the same way. Let $l_1 \in \{2, 3, ..., n\}$ be the minimal index for which $supp\, z_1^n \cap B_{l_1}^n \neq \emptyset$. The function $z_{l_1}^n$ is constant on $B_{l_1}^n$. Put

$$v_1 = z_1^n - \frac{z_1^n(t_{l_1})}{z_{l_1}^n(t_{l_1})} z_{l_1}^n,$$

where t_{l_1} is an arbitrary inner point of $B_{l_1}^n$. It is clear that

$$\int_{B_{l_1}^n} |v_1(t)| \, dt \geq \frac{1}{4},$$

and $supp\, v_1 \cap B_l^n = \emptyset$ for $l \in \{1, 2, ..., l_1\}$. If $l_1 < n$, then we put:

$$l_2 := \min\{l \in \{l_1 + 1, ..., n\} : supp\, v_1 \cap B_l^n \neq \emptyset\},$$

$$v_2 := v_1 - \frac{v_1(t_{l_2})}{z_{l_2}^n(t_{l_2})} z_{l_2}^n,$$

where t_{l_2} is an arbitrary inner point of $B_{l_2}^n$. We repeat the process until the set

$$\{l \in \{l_j + 1, ..., n\} : supp\, v_j \cap B_l^n \neq \emptyset\}, \quad 1 \leq j \leq n,$$

turns out to be empty. Then, we define $y_1^n := v_j$. By the construction, y_1^n satisfies (15.6) and (15.7).

It is clear that $[y_k^n]_{k=1}^n = [x_k]_{k=1}^n$, $\|y_k^n\|_{L_1} = 1$. Define the operator $P_n : L_1 \to [x_k]_{k=1}^n$ in the following way:

$$P_n(x) = \sum_{k=1}^{n} \frac{\int_{B_k^n} x(t)\,dt}{\int_{B_k^n} y_k^n(t)\,dt} y_k^n. \tag{15.8}$$

From (15.6) and (15.7) it follows that P_n is a projector on $[x_k]_1^n$ with $\|P_n\|_{L_1} \leq 4$.

In the above mentioned example we have:
for $k = 1, 2, 3, 4$,

$$y_k^6 = 4\kappa_{(\frac{k-1}{8},\frac{k}{8})} - 2\kappa_{(\frac{5}{8},\frac{3}{4})} - \kappa_{(\frac{3}{4},1)};$$

for $k = 5, 6$

$$y_k^6 = 8\kappa_{(\frac{1}{2}+\frac{k-5}{16},\frac{1}{2}+\frac{k-4}{16})} - 4\kappa_{(\frac{5}{8},\frac{3}{4})}.$$

Step 2. We are going to prove that the sequence $\{P_n x\}_{n=1}^\infty$ is a Cauchy one in L_1. Since the step functions with respect to dyadic intervals are dense in L_1, it suffices to check that the sequence $\{P_n \kappa_{\delta_i^j}\}_{n=1}^\infty$ be a Cauchy one for every $(i, j) \in \Omega$.

If $B_k^n \neq \{t : z_k^n(t) > 0\}$, $1 \leq k \leq n$, $n \in \mathbf{N}$, then, by Definition 15.3, for every $m > n$, there exists $k_m \in \{1, ..., m\}$ such that $B_k^n = B_{k_m}^m$. We shall call such sets stable. Let Q be the set of pairs (n, k), $n \in \mathbf{N}$, $1 \leq k \leq n$ such that B_k^n is stable, and also $B_k^n \notin \{B_i^l\}_{i=1}^l$ for $l \in \{1, ..., n-1\}$. Thus, if $(n, k) \in Q$, then there exists a sequence $\{k_m\}_{m=n}^\infty$, $1 \leq k_m \leq m$ for which $B_k^n = B_{k_m}^m$ for any $m \geq n$. Since $\{x_i\}_{i=1}^\infty$ satisfies Definition 15.3 and does not contain infinite bushes, it follows that the sequence $\{y_{k_m}^m\}_{m=n}^\infty$ is a Cauchy one in L_1. The sets B_k^n, $(n, k) \in Q$, are pairwise disjoint, that is why, for every $\varepsilon > 0$, there exists an $N_1 \in \mathbf{N}$ such that

$$\sum_{\{(n,k)\in Q:\ n\geq N_1\}} mes(B_k^n) < \varepsilon. \tag{15.9}$$

Since the number of pairs $(n, k) \in Q$ with $n \leq N_1$ is finite, it follows that there exists a number $N_2 > N_1$ with the following property. If the sequence $\{k_m\}_{m=n}^\infty$, $1 \leq k_m \leq m$, $(n, k) \in Q$, $n \leq N_1$ is such that $B_k^n = B_{k_m}^m$, $m \geq n$, then for every $p, q > N_2$

$$\|y_{k_p}^p - y_{k_q}^q\|_{L_1} < \varepsilon. \tag{15.10}$$

We also can assume that

$$mes(\cup_{l=1}^\infty supp\, x_l \setminus \cup_{m=1}^{N_1} supp\, x_m) < \varepsilon. \tag{15.11}$$

Let Q_1 be the set of pairs (n, k), $n \in \mathbf{N}$, $1 \leq k \leq n$, satisfying the following conditions: 1) $z_k^n \notin y_k^n$ on the set $\{t : z_k^n(t) < 0\}$; 2) $\{t : z_k^n(t) < 0\} \neq \{t : z_i^n(t) < 0\}$ for $i \in \{1, ..., k-1\}$; 3) for $l \in \{1, ..., n-1\}$, there exists no

$k_l \in \{1, ..., l\}$ for which $\{t : z_k^n(t) < 0\} = \{t : z_{k_l}^l(t) < 0\}$, and $z_{k_l}^l \neq y_{k_l}^l$ on the set $\{t : z_k^n(t) < 0\}$.

Since $\{x_n\}_1^\infty$ contains no infinite bush, we can assume that

$$mes(\cup_{l \in F} B_l^{N_1}) < \varepsilon, \tag{15.12}$$

where F is the set of such $l \in \{1, ..., N_1\}$, for which there exists no pair (m, k) with the following properties:

$$(m, k) \in Q_1, \quad m \leq N_1, \quad \{t : z_l^{N_1}(t) < 0\} = \{t : z_k^m(t) < 0\}.$$

By Definition 15.3, if $(n, k) \in \Omega_1$, then for every $m > n$ there exists a $k_m \in \{1, 2, ..., m\}$ such that $\{t : z_k^n(t) < 0\} = \{t : z_{k_m}^m(t) < 0\}$. Put $C_k^n := \{t : z_k^n(t) < 0\}, \quad (n, k) \in Q_1$. Since $\{x_i\}_1^\infty$ satisfies Definition 15.3 and contains no infinite bush, it follows that the sequence $\{y_{k_m}^m \kappa_{C_k^n}\}_{m=n}^\infty, (n, k) \in Q_1$, is a Cauchy one in L_1. The sets $D_k^n := \{t \in C_k^n : z_k^n(t) = y_k^n(t)\}$, $(n, k) \in Q_1$ are pairwise disjoint. That is why we can assume that

$$\sum_{\{(n,k) \in Q_1 : \ n \geq N_1\}} mes(D_k^n) < \varepsilon, \tag{15.13}$$

and for N_2 the following condition is satisfied. If $(n, k) \in Q_1$, $n \leq N_1$, and the sequence $\{k_m\}_{m=n}^\infty$, $k_m \in \{1, ..., m\}$ is such that $\{t : z_k^n(t) < 0\} = \{t : z_{k_m}^m(t) < 0\}$, $m \geq n$, then for every $p, q > N_2$

$$\|(y_{k_p}^p - y_{k_q}^q)\kappa_{C_k^n}\|_{L_1} < \varepsilon. \tag{15.14}$$

Let $q > p > N_2$. If $B_l^p = B_m^q = B_k^n$ for $(n, k) \in Q$, $n \leq N_1$, $1 \leq l \leq p$, $1 \leq m \leq q$, then by virtue of (15.6) and (15.10) we have that

$$\left\| \frac{mes(\Delta_i^j \cap B_l^p)}{\int_{B_l^p} y_l^p(t)\, dt} y_l^p - \frac{mes(\Delta_i^j \cap B_m^q)}{\int_{B_m^q} y_m^q(t)\, dt} y_m^q \right\|_{L_1} \leq 4mes(B_k^n)\varepsilon.$$

Let $\{t : z_l^p(t) < 0\} = \{t : z_k^n(t) < 0\}$ for $(n, k) \in Q_1$, $n \leq N_1$, $1 \leq l \leq p$. Put $S := \{s \in \{1, 2, ..., q\} : B_s^q \subset B_l^p\}$. By (15.6) and (15.14), we have that

$$\left\| \frac{mes(\Delta_i^j \cap B_l^p)}{\int_{B_l}^p y_l^p(t)\, dt} y_l^p - \sum_{s \in S} \frac{mes(\Delta_i^j \cap B_s^q)}{\int_{B_s}^q y_s^q(t)\, dt} y_s^q \right\|_{L_1} \leq$$

$$\leq 4mes(B_l^p)\varepsilon + 8 \sum_{\{(r,k) \in Q : \ r \geq N_1\}} mes(B_k^r \cap B_l^p).$$

By combining (15.6), (15.9), (15.11), (15,12), and (15,13), we get that for $q > p > N_2$

$$\|P_p(\kappa_{\Delta_i^j}) - P_q(\kappa_{\Delta_i^j})\|_{L_1} \leq 4mes(\cup_{l=1}^p supp\, x_l \backslash \cup_{m=1}^{N_1} supp\, x_m) +$$

$$+ 4mes(\cup_{l=1}^q supp\, x_l \backslash \cup_{m=1}^{N_1} supp\, x_m) + 8 \sum_{(n,k) \in Q : \ n \geq N_1} mes(B_k^n) +$$

$$+ 8 \sum_{(n,k) \in Q : \ n \geq N_1} mes(D_k^n) + 8\varepsilon + 2mes(\cup_{l \in F} B_l^{N_1}) < 34\varepsilon.$$

So, for every $x \in L_1$, the sequence $\{P_n x\}_1^\infty$ is a Cauchy one. It means that we can define the projector $Px := \lim_{n\to\infty} P_n x$ with $\|P\|_{L_1} \le 4$, $P : L_1 \to [x_k]_{k=1}^\infty$. Thus, Theorem 15.4 is proved for subsequences of the H.s. without any infinite bush.

Let $\{x_k\}_1^\infty = \{x'_k\}_{k=1}^\infty \cup \{T_l\}_{l\in L}$, where $\{x_k^1\}_{k=1}^\infty$ is a subsequence without any infinite bush, T_l is an infinite bush, L is a finite or an infinite set of indeces, and also, $\{x'_k\}_{k=1}^\infty$ and $\{T_l\}_{l\in L}$ have no common element. Let B_l be the largest support of the element from T_l, $l \in L$. Define the operator

$$Px = P'x + \sum_{l\in L}(x\kappa_{B_l} - \int_{B_l} x(t)\,dt),$$

where P' is a projector on $[x'_k]_{k=1}^\infty$. It is easy to show that P is a bounded projector on $[x_k]_1^\infty$. □

As an example, let us consider the subsequence of the H.s. $x_k = \chi_{2k}^1$; $k \in \mathbf{N}$. The natural projector on $[x_k]_1^\infty$ associated to the H.s. is unbounded. Indeed, for every $n \in N$,

$$\|\chi_0^1 + \sum_{l=1}^{2n} 2^{l-1}\chi_l^1\|_{L_1} = 1,$$

but

$$\|\sum_{k=1}^{n} 2^{2k-1}\chi_{2k}^1\|_{L_1} \ge \frac{n}{8}.$$

However, the subsequence $\{x_k\}_1^\infty$ is without branching, and that is why, by Theorem 15.4, the subspace $[x_k]_1^\infty$ is complemented in L_1. The projector $P : L_1 \to [x_k]_1^\infty$ can be defined in the explicit form

$$Px = 4\sum_{k=1}^{\infty}(\int_{A_k} x(t)\,dt)y_k,$$

where

$$A_k = (2^{-2k-2}, 2^{-2k-1}),$$

$$y_k = 2^{2k}(2\kappa_{(2^{-2k-3},2^{-2k-2})} + \kappa_{(2^{-2k-2},2^{-2k-1})} - \kappa_{(2^{-2k-1},2^{-2k})}).$$

Corollary 15.7. *The subspace of L_1 generated by a subsequence of the H.s. without branchings is isomorphic either to l_1 or to L_1.*

Now we want to investigate in detail the subsequences of the H.s. whose closed linear span in L_1 is isomorphic to l_1. The fact is, that there exist subsequences of the H.s., generating in L_1 the subspaces isomorphic to l_1, but which are not equivalent to the unit vector basis of l_1. For example, $[\chi_n^1]_{n=0}^\infty$ is isomorphic to l_1, but $\{\chi_n^1\}_{n=0}^{\infty}$ is not equivalent in L_1 to the unit vector basis of l_1 (see Remark 5.4). Let us find the conditions under which the subsequence of the H.s. is equivalent in L_1 to the unit vector basis of l_1.

Definition 15.8. *Let $\{d_n\}_1^{\infty}$ be a subsequence of the H.s. A sequence of sets $\{A_i\}_0^{\infty}$, $A_i \subset [0,1]$, is said to be a chain for $\{d_n\}_1^{\infty}$, if $A_{i+1} \subset A_i$, $i = 0,1,...,$ and there exist a sequence of signs $\{\varepsilon_j = \pm 1\}_1^{\infty}$ and a sequence of disjoint sets of natural numbers $\{N_i\}_0^{\infty}$ such that*

a) for every $n, m \in N_i$, $n \neq m$ $\quad supp\, x_n \cap supp\, x_m = \emptyset$;

b) $A_i = \{t : \sum_{j \in N_i} \varepsilon_j x_j(t) > 0\}$.

Lemma 15.9. *If a subsequence of the H.s. $\{d_n\}_1^{\infty}$ is unconditional in L_1, then there exist a natural number M and a non-negative scalar $C < 1$ such that an arbitrary chain $\{A_i\}_0^{\infty}$ for $\{d_n\}_1^{\infty}$ satisfies the condition*

$$mes(A_M) \leq C 2^{-M} mes(A_0). \tag{15.15}$$

Proof. Assume the contrary. Then, for every $M \in \mathbf{N}$ and every $\varepsilon > 0$, there exists a chain $\{A_i\}_0^{\infty}$ such that

$$mes(A_M) > (1-\varepsilon) 2^{-M} mes(A_0). \tag{15.16}$$

From Definition 15.8 it follows that

$$mes(A_i) \geq 2 mes(A_{i+1}) \tag{15.17}$$

for every $i = 0, 1, ...$ That is why, (15.16) and (15.17) imply that

$$mes(A_j) > (1-\varepsilon) 2^{-j} mes(A_0) \tag{15.18}$$

for every $j \in \{1, 2, ..., M\}$. Put $y_i := \sum_{j \in N_i} \varepsilon_j d_j$, $\quad i = 0, 1, ...,$ where $\{N_i\}_0^{\infty}$ and $\{\varepsilon_j\}_1^{\infty}$ have the properties described in Definition 15.8. We assume for convenience that M is even. Let us prove that

$$\|y_0 + \sum_{k=1}^{M} 2^{k-1} y_k\|_{L_1} \leq mes(A_0)(2 + M\varepsilon). \tag{15.19}$$

Indeed,

$$\begin{aligned}
&\int_0^1 \mid y_0(t) + \textstyle\sum_{k=1}^{M} 2^{k-1} y_k(t) \mid dt \leq \sum_{j=0}^{M-1} \int_{A_j \setminus A_{j+1}} \mid y_0(t) + \\
&+ \textstyle\sum_{k=1}^{M} 2^{k-1} y_k(t) \mid dt + \int_{[0,1] \setminus A_0} \mid y_0(t) \mid dt + \int_{A_M} \mid y_0(t) + \\
&+ \textstyle\sum_{k=1}^{M} 2^{k-1} y_k(t) \mid dt \leq \sum_{j=0}^{M-1} 2^j (mes(A_j) - 2 mes(A_{j+1})) + \\
&+ mes(A_0) + 2^M mes(A_M).
\end{aligned} \tag{15.20}$$

By combining (15.17) and (15.18), we have that

$$2^j(mes(A_j) - 2mes(A_{j+1})) \leq \varepsilon mes(A_0). \tag{15.21}$$

Inequality (15.19) follows from (15.20) and (15.21). On the other hand,

$$\| \sum_{k=1}^{M/2} 2^{2k-1} y_{2k} + y_0 \|_{L_1} \geq \sum_{j=0}^{M/2} \int_{A_{2j} \setminus A_{2j+1}} | y_0(t) +$$

$$+ \sum_{k=1}^{M/2} 2^{2k-1} y_{2k}(t) \mid dt \geq \sum_{j=1}^{M/2} \int_{A_{2j} \setminus A_{2j+1}} 2^{2j-1} \mid y_{2j}(t) \mid dt \geq$$

$$\geq \sum_{j=1}^{M/2} 2^{2j-2} mes(A_{2j}).$$

Together with (15.18) the last inequality implies that

$$\| \sum_{k=1}^{M/2} 2^{2k-1} y_{2k} + y_0 \|_{L_1} \geq (1-\varepsilon) 2^{-3} mes(A_0) M. \tag{15.22}$$

Since M and ε are arbitrary, inequalities (15.19) and (15.22) contradict the unconditionality of $\{d_n\}_1^\infty$. □

Lemma 15.10. *If there exist a natural number M and a non-negative scalar $c < 1$, such that every chain $\{A_i\}_0^\infty$ for the subsequence of the H.s. $\{d_n\}_1^\infty$ satisfies (15.15), then*

$$\| \max_{1 \leq i \leq n} \mid a_i d_i \mid \|_{L_1} \leq \frac{1}{1-c} 2^M \| \sum_{i=1}^n a_i d_i \|_{L_1} \tag{15.23}$$

for every $n \in \mathbf{N}$ and any scalars $\{a_i\}_1^n$.

Proof. To prove (15.23) it suffices to show that

$$mes(\{t : \max_{1 \leq i \leq n} |a_i d_i(t)| > \lambda\}) \leq \frac{1}{1-c} 2^M mes(\{t : | \sum_{i=1}^n a_i d_i(t)| > \lambda\}) \tag{15.24}$$

for every $\lambda > 0$.

Let $n_0(t) := \min\{i \in \{1, ..., n\} : |a_i d_i(t)| > \lambda\}$; N_0 is the set of values of $n_0(t)$, $t \in [0,1]$. Since $supp\, d_i \cap supp\, d_j = \emptyset$ for every $i, j \in N_0$, $i \neq j$, it follows that

$$mes(\{t : \max_{1 \leq i \leq n} |a_i d_i(t)| > \lambda\}) = \sum_{j \in N_0} mes(supp\, d_j). \tag{15.25}$$

Fix $j_0 \in N_0$. Note that the function $\sum_{i=1}^{j_0-1} a_i d_i$ is constant on $supp\, d_{j_0}$. We assume for definiteness that $\sum_{i=1}^{j_0-1} a_i d_i \geq 0$ on $supp\, d_{j_0}$, and $a_{j_0} > 0$ (the other cases are treated similarly). Then, we have that

$$\{t : supp\, d_{j_0} : |\sum_{i=1}^{j_0} a_i d_i(t)| > \lambda\} \supset \{t : d_{j_0}(t) = 1\}.$$

Define the chain for $\{d_i\}_1^n$:

$$A_0 := \{t : d_{j_0}(t) = 1\},$$

$$n_1(t) = \min\{i \in \{j_0 + 1, ..., n\} : \ t \in supp\, d_i \subset A_0\};$$

N_1 is the set of values of $n_1(t), \ t \in [0, 1]$;

$$A_1 = \{t : \textstyle\sum_{i \in N_1} a_i d_i(t) > 0\}.$$

The construction of the chain is continued by induction. If $\{n_j(t), N_j, A_j\}_{j=0}^k$ are defined, then we put

$$n_{k+1}(t) = \min\{i \in \{j_0 + 1, ..., n\} \setminus \cup_{j=0}^k N_j : t \in supp\, d_i \subset A_k\};$$

N_{k+1} is the set of values of $n_{k+1}(t)$;

$$A_{k+1} = \{t : \textstyle\sum_{i \in N_{k+1}} a_i d_i(t) > 0\}.$$

By condition (15.15), $mes(A_M) \leq c2^{-M} mes(A_0)$. That is why, we have that

$$mes(\{t \in supp\, d_{j_0} : |\textstyle\sum_{i=1}^n a_i d_i(t)| > \lambda\}) \geq$$

$$\geq mes(\{t \in supp\, d_{j_0} : |\textstyle\sum_{i=1}^{j_0} a_i d_i(t) +$$

$$+ \textstyle\sum_{i \in \cup_{j=1}^{M-1} N_j} a_i d_i(t)| > \lambda\} \setminus supp\, (\textstyle\sum_{i \in N_M} a_i d_i)) \geq 2^{1-M} mes(A_0) -$$

$$-c 2^{1-M} mes(A_0) = (1 - c) 2^{-M} mes(supp\, d_{j_0}).$$

Together with (15.25), the last inequality implies (15.24). Hence, (15.23) is proved. □

Now we can formulate the main result.

Theorem 15.11. *Let $\{d_n\}_1^\infty$ be a subsequence of the H.s. The following properties are equivalent:*

1) the sequence $\{d_n / \|d_n\|_{L_1}\}_1^\infty$ is equivalent to the unit vector basis of l_1;

2) $\{d_n\}_1^\infty$ is an unconditional basic sequence in L_1;

3) there exist $M \in \mathbf{N}$ and a non-negative scalar $c < 1$ such that every chain $\{A_i\}_0^\infty$ for $\{d_n\}_1^\infty$ satisfies (15.15).

Proof. It is obvious that 1) implies 2). By Lemma 15.9, 3) follows from 2). Let us prove that 3) implies 1). Put

$$m_0(t) := \min\{n : t \in supp\, d_n\};$$

N_0 is the set of values of $m_0(t)$, $t \in [0,1]$. It is clear that $supp\, d_i \cap supp\, d_j = \emptyset$ for every $i, j \in N_0$, $i \neq j$. Fix $i_0 \in N_0$. For $i = 1, 2$ we define

$$A_{0,i} := \{t : d_{i_0}(t) = (-1)^{i+1}\},$$

$$n_{1,i}(t) := \min\{n : t \in supp\, d_n \subset A_{0,i}\},$$

$N_{1,i}$ is the set of values of $n_{1,i}(t)$.

For $l = 1, 2, 3, 4$, we put

$$A_{1,l} = \{t : \sum_{j \in N_{1,[(l+1)/2]}} d_j(t) = (-1)^{l+1}\},$$

where $[a]$ denotes the standard "greatest integer function" of a. The process is continued by induction. If $\{n_{k,i}; N_{k,i}\}_{i=1}^{2^k}$ and $\{A_{k,l}\}_{l=1}^{2^{k+1}}$ are defined, then, for $j \in \{1, 2, ..., 2^{k+1}\}$, we put

$$n_{k+1,j}(t) := \min\{n : t \in supp\, d_n \subset A_{k,j}\},$$

$N_{k+1,j}$ is the set of values of $n_{k+1,j}(t)$.

For $l \in \{1, 2, ..., 2^{k+2}\}$ we define

$$A_{k+1,l} := \{t : \sum_{m \in N_{k+1,[(l+1)/2]}} d_m(t) = (-1)^{l+1}\}.$$

Note that $n_{k,j}$, $N_{k,j}$, $A_{k,l}$ depend on i_0, but for simplicity of notation we do not include this index in enumeration. By construction, $\{A_{k,p_k}\}_{k=0}^{\infty}$ is a chain for $\{d_n\}_1^{\infty}$, if

$$\left[\frac{p_{k+1}+1}{2}\right] = p_k. \tag{15.26}$$

Condition (15.15) implies that, for every sequence $\{p_k\}_{k=0}^{M}$ satisfying (15.26), there exists a k such that

$$mes(A_{k+1,p_{k+1}}) \leq 2^{-1} c^{1/M} mes(A_{k,p_k}). \tag{15.27}$$

For every sequence $\{p_k\}_{k=0}^{M}$ satisfying (15.26), let us choose the pair (k, p_k) with the minimal index k for which (15.27) is valid. The set of distinct chosen pairs is enumerated in an arbitrary way: $\{(k_j, p_{k_j})\}_{j=1}^{r}$, $2 \leq r \leq 2^M$. Put

$$B_j = A_{k_j,p_{k_j}} \setminus supp(\sum_{m \in N_{k_j+1,p_{k_j}}} d_m), \quad j = 1, ..., r.$$

Let $C_{i_0,j}$ be a subset of B_j, $j = 1, ..., r$, with the following property:

$$mes(C_{i_0,j} \cap supp\, d_q) = \frac{1}{M} mes(B_j \cap supp\, d_q) \tag{15.28}$$

for every $q \in N_{k_j,[(p_{k_j}+1)/2]}$. Since k_j is the minimal index, satisfying (15.27), it follows that

$$\begin{aligned} &mes(C_{i_0,j}) \geq \tfrac{1-c^{1/M}}{M} mes(A_{k_j,p_{k_j}}) \geq \\ &\geq (1-c^{1/M})c2^{-k_j} M^{-1} mes(A_{0,p_0}), \quad j = 1, ..., r. \end{aligned} \tag{15.29}$$

Furthermore, it is clear that

$$\sum_{j=1}^{r} 2^{-k_j} = 1.$$

That is why (15.29) implies that, for $D_{i_0} := \cup_{j=1}^{r} C_{i_0,j}$, we have the inequality

$$\int_{D_{i_0}} d_{i_0}(t)/\|d_{i_0}\|_{L_1}\, dt \geq M^{-1}(1-c^{1/M})c. \tag{15.30}$$

In a similar way we define D_i for every $i \in N_0$. It is obvious that $D_i \cap D_j = 0$ for distinct i and j from N_0. Let now $m_1(t) = \min\{n \in N_0 : t \in supp\, d_n\}$, N_1 is the set of values of $m_1(t)$. Using the same procedure it is possible, for $i \in N_1$, to define the sets D_i satisfying (15.30). By (15.28), we may require, in addition, that $D_i \cap D_j = \emptyset$ for every $i \in N_1$ and every $j \in N_0$. The process is continued by induction. If the sets D_i are constructed for $i \in \cup_{j=0}^{q} N_j$, $q \in \mathbf{N}$, then we put

$$m_{q+1}(t) = min\{n \notin \cup_{j=0}^{q} N_j : t \in supp\, d_n\},$$

N_{q+1} is the set of values of $m_{q+1}(t)$. Using the above described procedure, the sets D_i satisfying (15.30) are defined for $i \in N_{q+1}$ in such a way that $D_i \cap D_l = \emptyset$ for every $i \in N_{q+1}$ and every $l \in \cup_{j=0}^{q} N_j$. Finally, we obtain the sequence of pairwise disjoint sets $\{D_i\}_1^\infty$ for which (15.30) is valid. By Lemma 15.10 we get that

$$\begin{aligned} &\|\textstyle\sum_{i=1}^{n} a_i d_i/\|d_i\|_{L_1}\|_{L_1} \geq (1-c)2^{-M}\|\max_{1\leq i\leq n}|a_i d_i|/\|d_i\|_{L_1}\|_{L_1} \geq \\ &(1-c)2^{-M}\textstyle\sum_{i=1}^{n}\int_{D_i}|a_i d_i(t)|/\|d_i\|_{L_1}\, dt \geq \\ &\geq (1-c)(1-c^{1/M})cM^{-1}2^{-M}\textstyle\sum_{i=1}^{n}|a_i|, \quad n \in \mathbf{N}. \end{aligned}$$

The last inequality means that the sequence $\{d_i/\|d_i\|_{L_1}\}_1^\infty$ is equivalent to the unit vector basis of l_1. □

As it is shown in [30], there exists a subspace of L_1 isomorphic to l_1, but not complemented in L_1. Such situation is impossible in the case when the subspace is generated by a subsequence of the H.s. which is equivalent to the unit vector basis of l_1.

Theorem 15.12. *Let $\{d_n\}_1^\infty$ be a subsequence of the H.s. If $\{d_n/\|d_n\|_{L_1}\}_1^\infty$ is equivalent to the unit vector basis of l_1, then $[d_i]_1^\infty$ is complemented in L_1.*

Proof. By Lemma 15.9, there exist an $M \in \mathbf{N}$ and a non-negative $c < 1$ such that every chain $\{A_k\}_0^\infty$ for $\{d_i\}_1^\infty$ satisfies (15.15). Let us prove that, for every $n \in N$, there exists a projector $P_n : L_1 \to [d_i]_{i=1}^n$ with $\|P_n\|_{L_1} \le M(1 - c^{1/M})^{-M} c^{-1}$. It is proved in [79] that in such a situation $[d_i]_{i=1}^\infty$ is complemented in L_1.

Fix $n \in \mathbf{N}$. Let I_0 be the set of indeces $i \in \{1, ..., n\}$ for which there exists no $j \in \{1, ..., n\}\setminus\{i\}$ such that $supp\, d_j \subset supp\, d_i$. If $\{I_k\}_{k=0}^l$, $0 \le l < n$, have already been defined, then we denote by I_{l+1} the set of indeces $i \in \{1, ..., n\}\setminus \cup_{k=0}^l I_k$ for which there exists no $j \in \{1, ..., n\}\setminus((\cup_{k=0}^l I_k) \cup \{i\})$ such that $supp\, d_j \subset supp\, d_i$. Let r be the minimal index for which $I_{r+1} = \emptyset$. Then $\{1, ..., n\} = \cup_{k=0}^r I_k$. For $i \in I_0$ we put $m_i = M$, and for $k = 1, ..., m_i$ we define $e_{ik}^+ \subset \{t : d_i(t) > 0\}$ and $e_{ik}^- \subset \{t : d_i(t) < 0\}$ in such a way that, for $k \ne l$, $e_{ik}^+ \cap e_{il}^+ = e_{ik}^- \cap e_{il}^- = \emptyset$, and the following equations are valid:

$$mes(e_{ik}^+) = mes(e_{ik}^-) = c^{1/M}(1 - c^{1/M})(2M)^{-1} mes(supp\, d_i)$$

for $k = 1, ..., m_i$.

Let us consider $i \in I_1$. Two cases are possible.

I. At least one of the two inequalities is valid:

$$mes(\cup_{\{j \in \{1,...,n\}: supp\, d_j \subset \{t: d_i(t) > 0\}\}} supp\, d_j) > 2^{-1} c^{1/M} mes(supp\, d_i),$$

$$mes(\cup_{\{j \in \{1,...,n\}: supp\, d_j \subset \{t: d_i(t) < 0\}\}} supp\, d_j) > 2^{-1} c^{1/M} mes(supp\, d_i).$$

We assume for definiteness that the first one is valid. Then, we put $m_i = M - 1$, and for $k = 1, ..., m_i$ we define the sets

$$e_{ik}^+ \subset \cup_{\{j \in I_0:\ supp\, d_j \subset \{t: d_i(t) > 0\}\}} (e_{j,k+1}^+ \cup e_{j,k+1}^-)$$

with $mes(e_{ik}^+) = (1 - c^{1/M})^2 c^{2/M} (2M)^{-1} mes(supp\, d_i)$ in such a way that, for every $j \in I_0$,

$$m(e_{ik}^+ \cap e_{j,k+1}^+) = mes(e_{ik}^+ \cap e_{j,k+1}^-).$$

II. The following inequality is valid:

$$\max_{p=1,2}\{mes(\cup_{\{j \in \{1,...,n\}:\ supp\, d_j \subset \{t: d_i(t) = (-1)^p\}\}} supp\, d_j)\} \le$$

$$\le c^{1/M} 2^{-1} mes(supp\, d_i).$$

In this case we put $m_i = M$, and, for $k = 1, ..., m_i$, we define the sets

$$e_{ik}^{+} \subset \{t : d_i(t) > 0\} \backslash (\cup_{\{j=1,...,n:\ supp\, d_j \subset \{t: d_i(t)>0\}\}} supp\, d_j) \qquad (15.31)$$

satisfying the following conditions: if $k \neq l$, then $e_{ik}^{+} \cap e_{il}^{+} = \emptyset$; and

$$mes(e_{ik}^{+}) = (1 - c^{1/M})c^{1/M}(2M)^{-1} mes(supp\, d_i). \qquad (15.32)$$

The sets e_{ik}^{-}, for $k = 1, ..., m_i$ are defined in a similar way.

The process is continued by induction. Let the numbers $m_i \in \{1, ..., M\}$ and the sets $e_{ik}^{+} \subset \{t : d_i(t) > 0\}$, $e_{ik}^{-} \subset \{t : d_i(t) < 0\}$, be defined for $i \in \cup_{p=0}^{l} I_p$, $1 \leq l < r$, $1 \leq k \leq m_i$. And also, if $k \neq k_1$, then $e_{ik}^{+} \cap e_{ik_1}^{+} = e_{ik}^{-} \cap e_{ik_1}^{-} = \emptyset$; and $m(e_{ik}^{+}) = m(e_{ik}^{-}) = (1 - c^{1/M})^{M-m_i+1} c^{(M-m_i+1)/M} \times 2M^{-1} mes(supp\, d_i)$. For $i \in I_{l+1}$ two cases are possible.

I. At least one of the two inequalities is valid:

$$mes(\cup_{\{j \in \{1,...,n\}: supp\, d_j \subset \{t: d_i(t)>0\}\}} supp\, d_j) > 2^{-1} c^{1/M} mes(supp\, d_i),$$

$$mes(\cup_{\{j \in \{1,...,n\}: supp\, d_j \subset \{t: d_i(t)<0\}\}} supp\, d_j) > 2^{-1} c^{1/M} mes(supp\, d_i).$$

We assume for definiteness that the first one is valid. Denote by A_i the set of indeces with the following properties:

a) if $j_1 \neq j_2$; $j_1, j_2 \in A_i$, then $supp\, d_{j_1} \cap supp\, d_{j_2} = \emptyset$;

b) $\cup_{j \in A_i} supp\, d_j = \cup_{\{s \in \{1,...,n\}: supp\, d_s \subset \{t: d_i(t)>0\}\}} supp\, d_s$;

c) A_i is the minimal (with respect to the number of elements) set satisfying a) and b).

Put $A_{im} := \{j \in A_i : m_j = m\}$, $m = 1, ..., M$. By (15.15), we have that

$$mes(\cup_{j \in A_{i1}} supp\, d_j) \leq 2^{-1} c^{1/M} mes(supp\, d_i).$$

Let q be the maximal index for which

$$mes(\cup_{j \in \cup_{m=1}^{q} A_{im}} supp\, d_j) \leq 2^{-1} c^{1/M} mes(supp\, d_i).$$

Then we put $m_i = q$; and, for $k = 1, ..., m_i$, we define

$$e_{ik}^{+} \subset \cup_{j \in \cup_{m=q+1}^{M} A_{im}} (e_{j,k+1}^{+} \cup e_{j,k+1}^{-}) \cup (\{t : d_i(t) > 0\} \backslash \cup_{j \in A_i} supp\, d_j)$$

with

$$mes(e_{ik}^{+}) = (1 - c^{1/M})^{M-m_i+1} c^{(M-m_i+1)/M} (2M)^{-1} mes(supp\, d_i).$$

And also, if $k \neq l$, then $e_{ik}^{+} \cap e_{il}^{+} = \emptyset$; and $mes(e_{ik}^{+} \cap e_{j,k+1}^{+}) = mes(e_{ik}^{+} \cap e_{j,k+1}^{-})$ for every $j \in \cup_{p=0}^{l} I_p$.

II. If

$$\max_{p=1,2}\{mes(\cup_{\{j \in \{1,...,n\}: supp\, d_j \subset \{t: d_i(t)=(-1)^p\}\}} supp\, d_j)\} \leq$$

$$\leq 2^{-1} c^{1/M} mes(supp\, d_i),$$

then we put $m_i = M$; and, for $k = 1, ..., m_i$, we define the sets e_{ik}^{+} satisfying (15.31) and (15.32).

The sets e_{ik}^{-} are defined in a similar way. The process is stopped after the consideration of I_r.

Let $g_i = \kappa_{e_{i1}^{+}} - \kappa_{e_{i1}^{-}}$, $1 \le i \le n$. By construction, if $i \neq j$; $i, j \in \{1, ..., n\}$, then

$$\int_0^1 d_j(t) g_i(t)\, dt = 0.$$

Furthermore,

$$\int_0^1 \frac{d_i(t)}{\|d_i\|_{L_1}} g_i(t)\, dt \ge M^{-1}(1 - c^{1/M})^M c.$$

That is why the operator

$$P_n f := \sum_{i=1}^{n} \Big(\int_0^1 f(t) g_i(t)\, dt\Big)\Big(\int_0^1 d_i(t) g_i\, dt\Big)^{-1} d_i$$

is a projector on $[d_i]_1^n$. Since the sets $\{e_{i1}^{+}, e_{i1}^{-}\}_{i=1}^{n}$ are mutually disjoint, it follows that

$$\|P_n\|_{L_1} \le M(1 - c^{1/M})^{-M} c^{-1}.$$

Taking into account the remark made at the begining of the proof, Theorem 15.12 is proved. □

Chapter 16

Criterion of Equivalence of the Haar and Franklin Systems in R.I. Spaces

16.a. Definition and Basic Properties of the Franklin System

The Franklin system $\{f_m\}_{m=0}^{\infty}$ is an orthonormal system obtained by the Gram-Schmidt orthonormalization from the Faber-Schauder system $\{h_m\}_0^{\infty}$. Recall that

$$h_0(t) = \kappa_{[0,1]}(t),$$

$$h_m(t) = \int_0^t \chi_m(s)\,ds, \quad t \in [0,1], \quad m \in \mathbf{N}.$$

The basic properties of the Franklin system are the following:

a) $f_0 = \kappa_{[0,1]}, f_1(t) = \sqrt{3}(2t-1), \quad t \in [0,1]$;

b) if $m = 2^n + k$, for $(n,k) \in \Omega_1$, then f_m is a piecewise linear continuous function with the following points of break:

$$t_l^{(m)} = \begin{cases} l2^{-(n+1)}, & l = 0,1,...,2k; \\ (l-k)2^{-n}, l = 2k+1,...,m; \end{cases} \tag{16.a.1}$$

c) the following inequalities are valid

$$c_1 2^{n/2} q^{|2k-1-l|} \le (-1)^{l+1} f_m(t_l^{(m)}) \le c_2 2^{n/2} q^{|2k-1-l|}, \tag{16.a.2}$$

where $m = 2^n + k$, $(n,k) \in \Omega_1$, $l = 0,1,\dots,m$, $c_1 = (2+\sqrt{3})(3\sqrt{3})^{-1}$, $c_2 = 4\sqrt{3}(2+\sqrt{3})$, $q = (2+\sqrt{3})^{-1}$. In this section it is more convenient for us to normalize the Haar functions in L_2 and to use dyadic intervals for indexing. Let Q be the set of dyadic intervals Δ_n^k, $(n,k) \in \Omega_1$. If $I \in Q$, then the Haar function $\tilde{\chi}_I$ is defined as

$$\tilde{\chi}_I = | I |^{-1/2} (\kappa_{I'} - \kappa_{I''}),$$

where $|I| := mes(I)$; $I', I'' \in Q$, $I = I' \cup I'' = \emptyset$. It should be noted that only the first Haar function $\chi_1 = \kappa_{[0,1]}$ can not be written in such form.

The main result of this section is the following

Theorem 16.a.1. *The system of Haar* $\{\tilde{\chi}_m\}_{m=1}^{\infty}$ *and that of Franklin* $\{f_l\}_{l=0}^{\infty}$ *are equivalent in a separable r.i. space* E *if and only if* $0 < \alpha_E \le \beta_E < 1$.

The sufficiency of the condition $0 < \alpha_E \le \beta_E < 1$ will be proved in Section 16.d (see Corollary 16.d.3) after the demonstration of auxiliary results in Sections 16.b and 16.c. The necessity of the condition $0 < \alpha_E \le \beta < 1$ is shown in Section 16.e.

16.b. Martingale Transforms of the Haar Functions

The martingale transform of the martingale-difference sequence $\{d_n\}_1^{\infty}$ is the operator $T_\varepsilon : [d_n]_1^{\infty} \to [d_n]_1^{\infty}$ defined by equations $T_\varepsilon d_n = \varepsilon_n d_n$, $n \in \mathbf{N}$, where $\{\varepsilon_n\}_1^{\infty}$ is a sequence of scalars with $|\varepsilon_n| \le 1$ for every $n \in \mathbf{N}$. By (8.3), we have, for $p \in (1, \infty)$, that

$$\|T_\varepsilon\|_{L_p} \le (p^* - 1), \tag{16.b.1}$$

where $p^* = \max(p, p/(p-1))$.

Definition 16.b.1. *Let* $\tilde{Q}$ *be a finite set of indices,* $\{x_r\}_{r\in\tilde{Q}}$ *and* $\{y_r\}_{r\in\tilde{Q}}$ *are sequences of step functions. The sequence* $\{y_r\}_{r\in\tilde{Q}}$ *is said to be 1-dominated by* $\{x_r\}_{r\in\tilde{Q}}$ *if there exists a martingale transform which maps* x_r *into* y_r *for* $r \in \tilde{Q}$. *The notation of k-domination for* $k > 1$ *is defined by induction. The sequence* $\{y_r\}_{r\in\tilde{Q}}$ *is said to be k-dominated by* $\{x_r\}_{r\in\tilde{Q}}$ *if there exists a sequence* $\{z_r\}_{r\in\tilde{Q}}$ *such that* $\{y_r\}_{r\in\tilde{Q}}$ *is 1-dominated by* $\{z_r\}_{r\in\tilde{Q}}$ *and* $\{z_r\}_{r\in\tilde{Q}}$ *is* $(k-1)$*-dominated by* $\{x_r\}_{r\in\tilde{Q}}$. *If* $\tilde{Q}$ *is the infinite set of indeces, then* $\{y_r\}_{r\in\tilde{Q}}$ *is said to be k-dominated by* $\{x_r\}_{r\in\tilde{Q}}$, *if, for every finite subset* $Q' \subset \tilde{Q}$, $\{y_r\}_{r\in Q'}$ *is k-dominated by* $\{x_r\}_{r\in Q'}$.

The case, when one of the sequences is the sequence of the Haar functions, is the most interesting for us.

Definition 16.b.2. *Let* $\{h_I\}_{I\in\tilde{Q}}$ *be a sequence of step functions,* $\tilde{Q} \subset Q$. *The sequence* $\{h_I\}_{I\in\tilde{Q}}$ *is said to admit an m-splitting of degree k if there exist mutually disjoint sets* $Q_1, \ldots, Q_m$ *such that* $\tilde{Q} = \cup_{j=1}^{m} Q_j$ *and the sequence* $\{h_I\}_{I\in Q_j}$ *is k-dominated by* $\{\tilde{\chi}_I\}_{I\in Q_j}$ *for every* $j = 1, \ldots, m$.

The significance of the last notion is shown by the following

Lemma 16.b.3. *Let* $1 < p < \infty$, $p^* = p/(p-1)$, $\tilde{Q} \subset Q$; $m, k \in \mathbf{N}$. *If* $\{h_I\}_{I\in\tilde{Q}}$ *admits an m-splitting of degree k, then there exists a continuous*

linear operator H on L_p such that $H(\kappa_{[0,1]}) = 0$, $H(\tilde{\chi}_I) = h_I$ for every $I \in \tilde{Q}$, and

$$\|H\|_{L_p} \leq m(p^* - 1)^k p^*/2. \tag{16.b.2}$$

More exactly, for $p \in (1, 2]$,

$$\|H\|_{L_p} \leq c(p)(p^* - 1)^k \frac{p^*}{2} m^{1/2}, \tag{16.b.3}$$

and, for $p \in (2, \infty)$,

$$\|H\|_{L_p} \leq c(p)(p - 1)^k \frac{p}{2} m^{1-\frac{1}{p}}. \tag{16.b.4}$$

Proof. Let $Q_1, ..., Q_m$ be the decomposition of $\tilde{Q}$ such that $\{h_I\}_{I \in Q_j}$ is k-dominated by $\{\tilde{\chi}_I\}_{I \in Q_j}$ for $j = 1, ..., m$. By (16.b.1), for every sequence of scalars $\{c_I\}_{I \in Q_j}$ and for every $j = 1, ..., m$, we have that

$$\| \sum_{I \in Q_j} c_I h_I \|_{L_p} \leq (p^* - 1)^k \| \sum_{I \in Q_j} c_I \tilde{\chi}_I \|_{L_p}. \tag{16.b.5}$$

From (16.b.5) and (5.13) it follows that, for every sequence $\{c_I\}_{I \in Q}$,

$$\| \sum_{I \in \tilde{Q}} c_I h_I \|_{L_p} \leq \sum_{j=1}^{m} \| \sum_{I \in Q_j} c_I h_I \|_{L_p} \leq (p^* - 1)^k \sum_{j=1}^{m} \| \sum_{I \in Q_j} c_I \tilde{\chi}_I \|_{L_p} \leq$$

$$\leq (p^* - 1)^k \frac{p^*}{2} m \| \sum_{I \in Q} c_I \tilde{\chi}_I \|_{L_p}. \tag{16.b.6}$$

Inequality (16.b.6) means that the operator H, defined by equations $H(\kappa_{[0,1]}) = 0$, $H(\tilde{\chi}_I) = h_I$ for $I \in \tilde{Q}$, $H(\tilde{\chi}_I) = 0$ for $I \in Q \backslash \tilde{Q}$, satisfies inequality (16.b.2).

If $p \leq 2$, then, using (6.18), the Holder inequality, and 2-concavity of L_p, we get that

$$\| \sum_{I \in \tilde{Q}} c_I h_I \|_{L_p} \leq (p^* - 1)^k \sum_{j=1}^{m} \| \sum_{I \in Q_j} c_I \tilde{\chi}_I \|_{L_p} \leq$$

$$\leq (p^* - 1)^k c_2(L_p) \sum_{j=1}^{m} \| (\sum_{I \in Q_j} | c_I \tilde{\chi}_I |^2)^{1/2} \|_{L_p} \leq$$

$$\leq m^{1/2} (p^* - 1)^k c_2(L_p) (\sum_{j=1}^{m} \| (\sum_{I \in Q_j} | c_I \tilde{\chi}_I |^2)^{1/2} \|_{L_p}^2)^{1/2} \leq$$

$$\leq m^{1/2} (p^* - 1)^k c_2(L_p) \| (\sum_{j=1}^{m} \sum_{I \in Q_j} | c_I \tilde{\chi}_I |^2)^{1/2} \|_{L_p} \leq$$

$$\leq m^{1/2}(p^*-1)^k\frac{c_2(L_p)}{c_1(L_p)}\|\sum_{I\in\tilde{Q}}c_I\tilde{\chi_I}\|_{L_p}\leq$$

$$\leq m^{1/2}\frac{c_2(L_p)}{c_1(L_p)}(p^*-1)^k\frac{p^*}{2}\|\sum_{I\in Q}c_I\tilde{\chi_I}\|_{L_p}, \tag{16.b.7}$$

where $c_1(L_p)$, $c_2(L_p)$ are the constants from (6.18). Inequality (16.b.7) implies (16.b.3).

If $p > 2$, then it follows from p-concavity of L_p that

$$\|\sum_{I\in\tilde{Q}}c_Ih_I\|_{L_p}\leq(p-1)^kc_2(L_p)\sum_{j=1}^m\|(\sum_{I\in Q_j}|c_I\tilde{\chi_I}|^2)^{1/2}\|_{L_p}\leq$$

$$\leq c_2(L_p)(p-1)^km^{1-1/p}(\sum_{j=1}^m\|(\sum_{I\in Q_j}|c_I\tilde{\chi_I}|^2)^{1/2}\|_{L_p}^p)^{1/p}=$$

$$=c_2(L_p)(p-1)^km^{1-1/p}(\sum_{j=1}^m\int_0^1|\sum_{I\in Q_j}|c_I\tilde{\chi_I}(t)|^2|^{p/2}\,dt)^{1/p}\leq$$

$$\leq c_2(L_p)(p-1)^km^{1-1/p}(\int_0^1(\sum_{j=1}^m\sum_{I\in Q_j}|c_I\tilde{\chi_I}(t)|^2)^{p/2}\,dt)^{1/p}\leq$$

$$\leq\frac{c_2(L_p)}{c_1(L_p)}(p-1)^km^{1-1/p}\|\sum_{I\in\tilde{Q}}c_I\tilde{\chi_I}\|_{L_p}\leq$$

$$\leq\frac{c_2(L_p)}{c_1(L_p)}(p-1)^km^{1-1/p}\frac{p}{2}\|\sum_{I\in\tilde{Q}}c_I\tilde{\chi_I}\|_{L_p}.\quad\square$$

Remark 16.b.4. If E is an r.i. space with non-trivial Boyd indeces, then, under the assumptions of Lemma 16.b.3, there exists a continuous linear operator H on E such that $H(\kappa_{[0,1]})=0$, $\quad H(\tilde{\chi}_I)=h_I$ for every $I\in\tilde{Q}$, and

$$\|H\|_E\leq m(UC_E(\{\chi_i\}_1^\infty))^k\left(\frac{UC_E(\{\chi_i\}_1^\infty)+1}{2}\right).$$

Remark 16.b.5. Recall that a sequence $\{h_n\}_1^\infty$ is called block r-Hilbertian in E for $1\leq r<\infty$, if there exists a constant $c\in(0,\infty)$ such that

$$\|\sum_1^\infty c_nh_n\|_E\leq c(\sum_{j=1}^m\|\sum_{n\in N_j}c_nh_n\|_E^r)^{1/r} \tag{16.b.8}$$

for every sequence of scalars $\{c_n\}_1^\infty$ and for every finite partition $\mathbf{N} = \cup_{j=1}^\infty N_j$. If a sequence $\{h_I\}_{I\in\tilde{Q}}$ is block r-Hilbertian in L_p with $r = \min\{2, p\}$, then, under the assumptions of Lemma 16.b.3, we have that

$$\|H\|_{L_p} \le cc(p)(p^* - 1)^k \frac{p^*}{2} m^{\frac{1}{p}-\frac{1}{2}} \tag{16.b.9}$$

for $p \le 2$, and

$$\|H\|_{L_p} \le cc(p)(p - 1)^k \frac{p}{2} m^{\frac{1}{p}-\frac{1}{2}} \tag{16.b.10}$$

for $p > 2$.

Indeed, if (16.b.8) is valid, then, for $p \le 2$, we get that

$$\|\textstyle\sum_{I\in\tilde{Q}} c_I h_I\|_{L_p} \le c(\sum_{j=1}^m \|\sum_{I\in Q_j} c_I h_I\|_{L_p}^p)^{1/p} \le$$

$$\le c(p^* - 1)^k (\textstyle\sum_{j=1}^m \|\sum_{I\in Q_j} c_I \tilde{\chi}_I\|_{L_p}^p)^{1/p} \le$$

$$\le c(p^* - 1)^k [(\textstyle\sum_{j=1}^m \|\sum_{I\in Q_j} c_I \tilde{\chi}_I\|_{L_p}^2)^{p/2} m^{1-\frac{p}{2}}]^{1/p} \le$$

$$\le c\frac{c_2(L_p)}{c_1(L_p)}(p^* - 1)^k \frac{p^*}{2} m^{\frac{1}{p}-\frac{1}{2}} \|\textstyle\sum_{I\in Q} c_I \tilde{\chi}_I\|_{L_p}.$$

If $p > 2$, then

$$\|\textstyle\sum_{I\in\tilde{Q}} c_I\, h_I\|_{L_p} \le c(\sum_{j=1}^m \|\sum_{I\in Q_j} c_I\, h_I\|_{L_p}^2)^{1/2} \le$$

$$\le c\,(p-1)^k (\textstyle\sum_{j=1}^m \|\sum_{I\in Q_j} c_I\, \tilde{\chi}_I\|_{L_p}^2)^{1/2} \le$$

$$\le c\,(p-1)^k [(\textstyle\sum_{j=1}^m \|\sum_{I\in Q_j} c_I\, \tilde{\chi}_I\|_{L_p}^p)^{2/p} m^{1-2/p}]^{1/2} \le$$

$$\le c\,\frac{c_2(L_p)}{c_1(L_p)}\,(p-1)^k\, \frac{p}{2} m^{\frac{1}{2}-\frac{1}{p}}\, \|\textstyle\sum_{I\in Q} c_I\, \tilde{\chi}_I\|_{L_p}.$$

Remark 16.b.6. *If* $1 \le p < \infty$, $r = \min\{p, 2\}$, *then every unconditional basic sequence* $\{h_I\}_{I\in\tilde{Q}}$ *is block* r*-Hilbertian in* L_p.

In fact, by Remark 1.b.9, we have that

$$c^{-1}\|(\sum_{I\in\tilde{Q}} |\, c_I h_I\,|^2)^{1/2}\|_{L_p} \le \|\sum_{Iin\tilde{Q}} c_I h_I\|_{L_p} \le$$

$$\le c\|(\sum_{I\in\tilde{Q}} |\, c_I h_I\,|^2)^{1/2}\|_{L_p}. \tag{16.b.11}$$

Let $\tilde{Q} = \cup_{j=1}^m Q_j$ be a finite partition of $\tilde{Q}$. Put $x_j = (\sum_{I\in Q_j} |c_I h_I|^2)^{1/2}$. Since $l_p \subset l_2$ for $p < 2$, we get that

$$\|(\sum_{j=1}^m |x_j|^2)^{1/2}\|_{L_p} = (\int_0^1 |\sum_{j=1}^m |x_j(t)|^2|^{p/2}\, dt)^{1/p} \le$$

$$\leq \left(\int_0^1 \sum_{j=1}^m |x_j(t)|^p \, dt\right)^{1/p} = \left(\sum_{j=1}^m \|x_j\|_{L_p}^p\right)^{1/p}. \tag{16.b.12}$$

From (16.b.11) and (16.b.12) it follows that

$$\|\sum_{I\in\tilde{Q}} c_I h_I\|_{L_p} \leq c(\sum_{j=1}^m \|(\sum_{I\in Q_j} (c_I h_I)^2)^{1/2}\|_{L_p}^p)^{1/p} \leq c^2(\sum_{j=1}^m \| \sum_{I in Q_j} c_I h_I\|_{L_p}^p)^{1/p}.$$

If $p > 2$, then

$$\begin{aligned} &\|(\textstyle\sum_{j=1}^m \mid x_j \mid^2)^{1/2}\|_{L_p} = \| \textstyle\sum_{j=1}^m \mid x_j \mid^2 \|_{L_{p/2}}^{1/2} \leq \\ &\leq (\textstyle\sum_{j=1}^m \| \mid x_j \mid^2 \|_{L_{p/2}})^{1/2} = (\textstyle\sum_{j=1}^m \|x_j\|_{L_p}^2)^{1/2}. \end{aligned} \tag{16.b.13}$$

Inequalities (16.b.11) and (16.b.13) imply (16.b.8). □

Let us describe the situations when a family of step functions is 1-dominated by another one.

Lemma 16.b.7. *Let $\{x_i\}_1^n$ and $\{y_i\}_1^n$ be sequences of step functions on [0,1] with the following property: for every $i = 1, ..., n$ there exist a martingale difference sequence $d_{i,1}, ..., d_{i,m(i)}$ and a sequence of scalars $\varepsilon_{i,1}, ..., \varepsilon_{i,m(i)}$ with $|\varepsilon_{ij}| \leq 1$, $m(i) \in \mathbf{N}$, such that*

$$x_i = \sum_{j=1}^{m(i)} d_{i,j}, \quad y_i = \sum_{j=1}^{m(i)} \varepsilon_{ij} d_{i,j}.$$

Assume also that, for every $i = 1, ..., n$, there exists a measurable set A_i such that

$$a) \ \mathrm{supp}\, d_{i,j} \subset A_i, j = 1, ..., m(i); \tag{16.b.14}$$

$$\begin{aligned} &b) \text{ the functions } d_{s,j} \text{ are constant on } A_i \\ &\text{for } s = 1, ..., i-1; \ j = 1, ..., m(i). \end{aligned} \tag{16.b.15}$$

Then the sequence $\{y_i\}_1^n$ is 1-dominated by $\{x_i\}_1^n$, i.e. there exists a martingale transform T such that $T(x_i) = y_i$ for $i = 1, ..., n$. If $(\varepsilon_{ij})^2 = 1$ for every i and j, then $T(y_i) = x_i$.

Proof. Introduce the lexicographical ordering of the set $L := \{(i,j) : 1 \leq i \leq n, \ \ 1 \leq j \leq m(i)\}$. This means that $(i_1, j_1) \prec (i_2, j_2)$ if either $i_1 < i_2$ or $i_1 = i_2$ and $j_1 \leq j_2$. If we enumerate the functions $\{d_{i,j}\}_{(i,j)\in L}$ according to this order, then from (16.b.14) and (16.b.15) it follows that $\{d_m\}_{m=1}^{cardL}$ is a martingale difference sequence. A martingale transform T, defined by equations $T(d_m) = \varepsilon_m d_m$, satisfies the assertion of Lemma 16.b.7. □

Let us investigate rearrangements of the H.s. with the help of the introduced notions. Let $k \in \mathbf{N}$, $I \in Q$, $\mid I \mid \leq 2^{-k}$. Denote by $I^{(k)}$ the interval $J \in Q$ such that $J \supset I$ and $\mid J \mid = 2^k \mid I \mid$.

Lemma 16.b.8. *Let π be a rearrangement of intervals $I \in Q$ such that $|\pi(I)| = |I|$ for every $I \in Q$.*

a) If $k \geq 1$ and $\pi(I) \subset I^{(k)}$ for every $I \in Q$, then the sequence $\{\tilde{\chi}_{\pi(I)}\}_{I\in Q}$ admits a $3(k+1)$-splitting of degree 1.

b) Let $k \geq 2$, and a map $\mu : Q \to \{0,1\}$ satisfies $\mu(I) = 0$ whenever the distance between $I^{(k)}$ and $(\pi(I))^{(k)}$ is not equal to 0. Then the sequence $\{\mu(I)\tilde{\chi}_{\pi(I)}\}_{I\in Q}$ admits a $27(k+1)$-splitting of degree 3.

Proof. Step 0. It is clear that the sequence

$$d_{I,1} = \frac{1}{2}(\tilde{\chi}_I + \tilde{\chi}_{\pi(I)}), \quad d_{I,2} = \frac{1}{2}(\tilde{\chi}_I - \tilde{\chi}_{\pi(I)})$$

is a martingale difference. Furthermore, a martingale transform

$$Td_{I,1} = d_{I,1}, \quad Td_{I,2} = -d_{I,2}$$

maps $\tilde{\chi}_I$ into $\tilde{\chi}_{\pi(I)}$.

Step 1. Let us prove first the formulated above statement a) under the additional assumption that $\pi(\pi(I)) = I$ for $I \in Q$. Define the function $\nu : Q \to \{0,1,...,k\}$ in the following way: $\nu(I) = r$ if $|I| = 2^{-n}$ and $(n-r)$ is divisible by $k+1$. Put

$$A(I) := I \cup \pi(I) \text{ for } I \in Q,$$

$$Q_r := \{I \in Q : \nu(I) = r\} \text{ for } r \in \{0,...,k\}.$$

Then the sequence $\{\tilde{\chi}_{\pi(I)}\}_{I\in Q_r}$ is 1-dominated by $\{\tilde{\chi}_I\}_{I\in Q_r}$. This follows from Lemma 16.b.7 if we use it with the martingale difference sequences defined in Step 0. Conditions (16.b.14) and (16.b.15) are fulfilled because of the inclusion: $A(I) \subset I^{(k)}$ for $I \in Q$.

Step 2. Let us prove statement a) in a general case. Let the function $\eta : Q \to \{0,1,2\}$ satisfy the condition:

$$\text{if } \eta(\pi(I)) = \eta(I), \text{ then } \pi(I) = I. \tag{16.b.16}$$

Such a function can be defined in the following way. Let $P(I) := \{\pi^i(I), i \in \mathrm{N}\}$ be the orbit of π. If $P(I)$ has n elements and n is even, then we put

$$\eta(\pi^i(I)) := \frac{1}{2}(1 + (-1)^i), \ i = 1,...,n.$$

If n is odd, then

$$\eta(\pi^i(I)) := \frac{1}{2}(1 + (-1)^i), \ i = 1,...,n-1,$$

$$\eta(\pi^n(I)) := 2.$$

The functions ν (see Step 1) and η define a $3(k+1)$-splitting of degree 1. To prove this we again use Lemma 16.b.7. Indeed, if $|I_1| \neq |I_2|$, then assumptions (16.b.14) and (16.b.15) are valid due to the same reason as in Step 1. If $|I_1| = |I_2|$, $I_1 \neq I_2$, and $\eta(I_1) = \eta(I_2)$, then, by (16.b.16), $I_1 \neq \pi(I_2)$, $I_2 \neq \pi(I_1)$. Thus, $A(I_1) \cap A(I_2) = \emptyset$, and conditions (16.b.14), (16.b.15) are fulfilled.

We pass now to the proof of statement b).

Step 3. Let Q^0 be the set of $I \in Q$ such that either $\pi(I) \subset I^{(k)}$ or $\mu(I) = 0$. It is clear that there exists a rearrangement π_1 coinciding with π on the intervals I with $\pi(I) \subset I^{(k)}$ and satisfying the assumptions of statement a). By Step 2 the sequence $\{\tilde{\chi}_{\pi_1(I)}\}_{I\in Q^0}$ admits a $3(k+1)$-splitting of degree 1. This implies that $\{\mu(I)\tilde{\chi}_{\pi(I)}\}_{I\in Q^0}$ admits a $3(k+1)$-splitting of degree 2.

Put $Q^1 = Q\backslash Q^0$. Let Q^2 be the set of $I \in Q$ such that the boundary of I is disjoint with the boundary of $I^{(k)}$. Put

$$Q^3 = \{I \in Q^1 : I \in Q^2,\ \pi(I) \in Q^2\}.$$

We are going to prove that $\{\tilde{\chi}_{\pi(I)}\}_{I\in Q^3}$ admits a $6(k+1)$-splitting of degree 1 (The factor $\mu(I)$ is omitted since $\mu(I) = 1$ for $I \in Q^1$). We shall use the functions ν and η defined in Step 1 and 2. Moreover, we can assume that the function η takes only two values 0 and 1 because the inclusion $P(I) \subset Q^1$ implies that the number of elements of $P(I)$ is even. We also need the function $\xi : Q^3 \to \{0,1,2\}$ satisfying the following condition:

(A) let $I_1, I_2 \in Q^3$, $\nu(I_1) = \nu(I_2)$, $\eta(I_1) = \eta(I_2)$, and $|I_1| > |I_2|$; if at least one of the functions $\tilde{\chi}_J$, $J \in \{I_1, \pi(I_1)\}$ is not constant on the set $A(I_2) = I_2 \cup \pi(I_2)$, then $\xi(I_1) \neq \xi(I_2)$.

It is clear that the functions ν, η and ξ define a $6(k+1)$-splitting of degree 1 for $\{\tilde{\chi}_I\}_{I\in Q^3}$. Conditions (16.b.14), (16.b.15) follow from (A).

Let us give the construction of the function ξ. Put $ord(\xi) = \inf\{m \in \mathbf{N} : 2^m\xi \in \mathbf{N}\}$, $\xi \in [0,1]$. We define the functions $label_{k,r}(\xi)$, $r = 0, 1, ..., k$ for ξ with $ord(\xi) < \infty$ by induction with respect to $ord(\xi)$. If $ord(\xi) \leq r+1$, then $label_{k,r}(\xi) := 0$. If $r + 1 < ord(\xi) < \infty$, then we find the shortest $I \in Q$ such that

$$\xi \in I,\ |I| > 2^{-ord(\xi)},\ \nu(I^{(1)}) = r. \qquad (16.b.17)$$

We put

$$label_{k,r}(\xi) := \min(\{0,1,2\}\backslash\{label_{k,r}(\inf I),\ label_{k,r}(\sup I)\}). \qquad (16.b.18)$$

The function $\xi : Q^3 \to \{0,1,2\}$ is defined by the formula

$$\xi(I) = label_{k,\nu(I)}(\xi_k(I)), \qquad (16.b.19)$$

where $\xi_k(I)$ is a unique common end-point of the intervals $I^{(k)}$ and $(\pi(I))^k$.

Let us prove that the function ξ satisfies condition (A). Let I_1, I_2, J comply with the assumptions of (A), $r := \nu(I_1) = \nu(I_2)$, $\lambda_1 := \xi_k(I_1)$, $\lambda_2 := \xi_k(I_2)$. Then $|J| \geq 2^{k+1}|I_2|$. If $\tilde{\chi}_J$ is not constant on $I_2 \cup \pi(I_2)$, then λ_2 is a point of discontinuity for $\tilde{\chi}_J$, since $I_2(k)$ and $(\pi(I_2))^{(k)}$ are the adjacent dyadic intervals and $|\, I_2^{(k)} \cup (\pi(I_2))^{(k)} \,| \leq |J|$. Furthermore, λ_2 is not an end-point of $J^{(k)}$ because $J \in Q^3 \subset Q^2$. The fact that λ_2 is a point of discontinuity for $\tilde{\chi}_J$ implies that

$$2^{-1}|J| \leq 2^{-ord(\lambda_2)} < |J^{(k)}| = 2^k|J|.$$

Consequently, $J^{(k)}$ is the shortest interval satisfying (16.b.17) for $\xi = \lambda_2$. Since λ_1 is equal either to $\inf J^{(k)}$ or to $\sup J^{(k)}$, it follows from (16.b.18) that $\xi(\lambda_2) \neq \xi(\lambda_1)$. Thus, the function ξ satisfies property (A).

Step 4. Let π_1 be the map which rearranges the intervals I_1 and I_2 whenever $I_1 \cup I_2 = I_1^{(1)} = I_2^{(1)}$. If $k \geq 2$, then $I \in Q^2$ implies $\pi_1(I) \in Q^2$ (remind that the set Q^2 is defined in terms of k). By Step 1, the sequence $\{\chi_{\pi_1(I)}\}_{I\in Q}$ admits a $(k+1)$-splitting of degree 1 defined by the function ν. Put

$$Q^4 = \{I \in Q^1 \backslash Q^3 : I \notin Q^2, \pi(I) \in Q^2\};$$

$$Q^5 = \{I \in Q^1 \backslash Q^3 : I \in Q^2, \pi(I) \notin Q^2\};$$

$$Q^6 = \{I \in Q^1 \backslash Q^3 : I \notin Q^2, \pi(I) \notin Q^2\}.$$

By Step 3, the sequence $\{\tilde{\chi}_{\pi\circ\pi_1(I)}\}_{I\in\pi_1(Q^4)}$ admits a $6(k+1)$-splitting of degree 1. It is clear that this splitting is a refinement of a $(k+1)$-splitting of order 1 for $\{\tilde{\chi}_{\pi_1(I)}\}_{I\in Q}$. Since $\pi = (\pi \circ \pi_1) \circ \pi_1$, the sequence $\{\tilde{\chi}_{\pi(I)}\}_{I\in Q^4}$ admits a $6(k+1)$-splitting of degree 2. The case of $\{\tilde{\chi}_{\pi(I)}\}_{I\in Q^5}$ is treated similarly using the formula $\pi = \pi_1 \circ (\pi_1 \circ \pi)$. For $\{\tilde{\chi}_{\pi(I)}\}_{I\in Q^6}$ we have a $6(k+1)$-splitting of degree 3 due to the equation $\pi = \pi_1 \circ (\pi_1 \circ \pi \circ \pi_1) \circ \pi_1$.

The obtained splittings of all parts of Q give a $27(k+1)$-splitting of degree 3 for the sequence $\{\mu(I)\tilde{\chi}_{\pi(I)}\}_{I\in Q}$. □

Remark 16.b.9. Lemmas 16.b.3 and 16.b.8 give sufficient conditions for boundedness in L_p, $1 < p < \infty$, of rearrangements of the H.s. This subject is presented also in Chapter 11.

16.c. Norm Estimates of Auxiliary Operators

Let us introduce, for integer values of l, the linear operators V_l and W_l by the formulas:

$$V_l\kappa_{[0,1]} = W_l\kappa_{[0,1]} = 0;$$

$$V_l(\tilde{\chi}_I) = \begin{cases} \tilde{\chi}_{I+l|I|}, & \text{if } I + l|I| \subset [0,1]; \\ 0, & \text{otherwise}; \end{cases}$$

$$W_l(\tilde{\chi}_I) = \begin{cases} |\,\tilde{\chi}_{I+l|I|}\,| - |\,\tilde{\chi}_I\,|, & \text{if } I + l|I| \subset [0,1]; \\ 0, & \text{otherwise}; \end{cases}$$

where $I \in Q$, $a + I := \{a + t : t \in I\}$, $a \in \mathbf{R}$.

Lemma 16.c.1. *Let* $1 < p < \infty$; $l, j \in \mathbf{Z}$, $j \geq 0$, *and let* $\{h_I\}_{I \in Q}$ *be a sequence of functions of the form*

$$h_I = \sum_{\{J:|J|=2^{-j}|I|\}} a_{I,J} \tilde{\chi}_J,$$

where the coefficients $a_{I,J}$ *are such that*

$$|h_I| \leq |V_l(\tilde{\chi}_I)|, \quad I \in Q. \tag{16.c.1}$$

Then the operator V, *defined by the equations* $V\kappa_{[0,1]} = 0$, $\quad V\tilde{\chi}_I = h_I$ *for* $I \in Q$, *satisfies the estimate*

$$\|V\|_{L_p} \leq c\|V_l\|_{L_p}, \tag{16.c.2}$$

where the constant c *depends only on* p.

Proof. It is clear that $\{h_I\}_{I \in Q}$ and $\{V_l(\tilde{\chi}_I)\}_{I \in Q}$ are martingale differences. That is why, by Corollary 8.11 and (16.c.1), we have that, for every sequence of scalars $\{c_I\}_{I \in Q}$,

$$\|\sum_{I \in Q} c_I h_I\|_{L_p} \leq c_2(p)\|(\sum_{I \in Q}(c_I h_I)^2)^{1/2}\|_{L_p} \leq$$

$$\leq c_2(p)\|(\sum_{I \in Q}(c_I V_l(\tilde{\chi}_I))^2)^{1/2}\|_{L_p} \leq \frac{c_2(p)}{c_1(p)}\|\sum_{I \in Q} c_I V_l(\tilde{\chi}_I)\|_{L_p},$$

where $c_1(p) := c_1(L_p)$ and $c_2(p) := c_2(L_p)$ are constants from (8.19). □

Remark 16.c.2. Lemma 16.c.1 is valid in an arbitrary r.i. space with non-trivial Boyd indices.

Theorem 16.c.3. *Let* $1 < p < \infty$. *Then the operators* V_l *and* W_l *are bounded in* L_p *for every* $l \in \mathbf{Z}$, *and the following estimates are fulfilled:*

$$\|V_l\|_{L_p} \leq c(p)(\log(2 + |l|))^{\frac{1}{p} - \frac{1}{2}}, \tag{16.c.3}$$

$$\|W_l\|_{L_p} \leq c(p)(\log(2 + |l|))^{\frac{1}{p} - \frac{1}{2}}, \tag{16.c.4}$$

for $p \leq 2$;

$$\|V_l\|_{L_p} \leq c(p)(\log(2 + |l|))^{\frac{1}{2} - \frac{1}{p}}, \tag{16.c.5}$$

$$\|W_l\|_{L_p} \leq c(p)(\log(2 + |l|))^{\frac{1}{2} - \frac{1}{p}}, \tag{16.c.6}$$

for $p > 2$.

Proof. To prove the norm estimates of the operator V_l for $l \in \mathbf{Z}$ we use Lemma 16.b.8 with the smallest $k \geq 2$ such that $|l| \leq 2^k$. Due to this lemma the sequence $\{V_l(\tilde{\chi}_I)\}_{I \in Q}$ admits a $27(k+1)$-splitting of degree 3. By Corollary 8.4. and Remark 16.b.6, the operator V_l satisfies inequalities (16.b.9) and (16.b.10). For $p \leq 2$ we have that

$$\|V_l\|_{L_p} \leq c(p)(p^* - 1)^3 \frac{p^*}{2}(27(k+1))^{\frac{1}{p}-\frac{1}{2}}.$$

This estimate clearly implies (16.c.3). If $p > 2$, then from (16.b.10) it follows that

$$\|V_l\|_{L_p} \leq c(p)(p-1)^3 \frac{p}{2}(27(k+1))^{\frac{1}{2}-\frac{1}{p}}.$$

The last inequality is equivalent to (16.c.5).

To prove the norm estimates of W_l for $l \in \mathbf{Z}$ it is necessary to modify the proof of Lemma 16.b.8. By reasons of symmetry, we may assume that $l > 0$. Let us consider first the case where $l > 1$. Let $k \in \mathbf{N}$ and $2^k > l$. Define the auxiliary rearrangement $\pi : Q \to Q$ in the following way:

$$\pi(I) = \begin{cases} I + l|I|, & \text{if } I + l|I| \subset [0,1], \\ I + l|I| - n, & \text{if } I + l|I| \ \subset [0,1], \end{cases}$$

where $n \in \mathbf{N}$ depends on $|I|$ and is chosen such that $\pi(I) \subset [0,1]$ for $I \in Q$. Let $\mu(I) = 1$, when $\pi(I) = I + l|I|$; and $\mu(I) = 0$ in the opposite case. The sets Q^0 and Q^3 are defined as in Lemma 16.b.8. In order to use Lemma 16.b.7 we have to replace the martingale difference sequence, used in Step 0 of the proof of Lemma 16.b.8, by the following one. Let $I, J, I', I'' \in Q$, $|I| = |J|$, $I \neq J$, $I' \cup I'' = I$, $I' \cap I'' = \emptyset$. Put

$$d_1 = |I|^{-1/2}\left(\frac{1}{3}\kappa_{I' \cup J} - \kappa_{I''}\right),$$

$$d_2 = |I|^{-1/2}\left(\frac{2}{3}\kappa_{I'} - \frac{1}{3}\kappa_J\right).$$

Then, $d_1 + d_2 = \tilde{\chi}_I$ and

$$\frac{1}{2}d_1 - d_2 = \frac{1}{2}(|\tilde{\chi}_J| - |\tilde{\chi}_I|).$$

Repeating the argument of Step 1-3 of the proof of Lemma 16.b.8 with the new martingale difference sequence, we obtain that $\{\frac{1}{2}W_l(\tilde{\chi}_I)\}_{I \in Q^0 \cup Q^3}$ admits a $9(k+1)$-splitting of degree 1. Remark now that, for $l < 2^k$,

$$Q \backslash (Q^0 \cup Q^3) = Q^7 \cup Q^8,$$

where

$$Q^7 := \{I \in Q \backslash Q^0 : \ \sup I = \sup I^{(k)} < 1\},$$

$$Q^8 := \{I \in Q \backslash Q^0 : \inf(I + l|I|) = \sup I^{(k)} < 1\}.$$

Assume, in addition, that $l < 2^k - 1$. Let $Q^9 = Q^8 \cup \{I \in Q : I + |I| \in Q^7\}$. The described above reasoning shows that the sequence $\{\frac{1}{2} W_{l+1} \tilde{\chi}_I\}_{I \in Q^9}$ admits a $9(k+1)$-splitting of degree 1.

In what follows we need the auxiliary operator D_1, defined by the conditions: $D_1(\tilde{\chi}_I) = \tilde{\chi}_{I^{(1)}}$, if $|I| \leq 2^{-1}$; and $D_1(\tilde{\chi}_{[0,1]}) = D_1(\kappa_{[0,1]}) = 0$. The operator D_1 is bounded in L_p. Indeed, it is clear that

$$D_1^*(\tilde{\chi}_J) = \tilde{\chi}_{J'} + \tilde{\chi}_{J''}, \quad D_1^*(\kappa_{[0,1]}) = 0,$$

where D_1^* is the adjoint operator; $J, J', J'' \in Q, \quad J' \cup J'' = J, \quad J' \cap J'' = \emptyset$. The operator D_1^* is continuous in L_p^* due to Lemma 16.c.1.

Now the continuity of $W_l \circ P_{Q^7}$ ($P_{\tilde{Q}}$ is the natural projector on $\{\tilde{\chi}_I\}_{I \in \tilde{Q}}$ associated to the H.s.) follows from the formula

$$W_{l+1} \circ V_{-1}(P_{Q^7}(\tilde{\chi}_I)) + \sqrt{2} D_1(P_{Q^7}(\tilde{\chi}_I)) = W_l(P_{Q^7}(\tilde{\chi}_I)).$$

Really, for every $I \in Q^7$, we have that

$$W_{l+1} \circ V_{-1}(\tilde{\chi}_I) + \sqrt{2} D_1(\tilde{\chi}_I)) =$$

$$= |\tilde{\chi}_{I+l|I|}| - |\tilde{\chi}_{I+l|I|}| - (|\tilde{\chi}_{I-|I|}| + |\tilde{\chi}_I|) = |\tilde{\chi}_{I+l|I|}| - |\tilde{\chi}_I| = W_l(\tilde{\chi}_I).$$

If we define $l' \in N$ by the condition: $l \leq 2l' < l + 2$, then

$$W_{l+1}(P_{Q^8}(\tilde{\chi}_I)) + \sqrt{2} V_{l'} D_1(P_{Q^8}(\tilde{\chi}_I)) = W_l(P_{Q^8}(\tilde{\chi}_I)).$$

In fact, for $I \in Q^8$, we get that

$$W_{l+1}(\tilde{\chi}_I) + \sqrt{2} V_{l'} D_1(\tilde{\chi}_I) = |\tilde{\chi}_{I+(l+1)|I|}| - |\tilde{\chi}_I| + \sqrt{2} \tilde{\chi}_{I^{(1)}+l'|I^{(1)}|} =$$

$$= |\tilde{\chi}_{I+(l+1)|I|}| - |\tilde{\chi}_I| + |\tilde{\chi}_{I+l|I|}| - |\tilde{\chi}_{I+(l-1)|I|}| = W_l(\tilde{\chi}_I).$$

This formula means that the operator $W_l \circ P_{Q^8}$ is bounded in L_p.

Thus, the operator W_l can be expressed as the sum of three operators

$$W_l = W_l \circ P_{Q^0 \cup Q^3} + W_l \circ P_{Q^7} + W_l \circ P_{Q^8},$$

which have the needed norm estimates due to Remark 16.b.5, where $k \in \mathbf{N}$ is chosen to satisfy the condition $2^k \geq l + 2 > 2^{k-1}$. So, inequalities (16.c.4) and (16.c.6) are proved for $l > 1$.

The case of $l = 1$ follows from the identity

$$W_1 = (W_2 + \sqrt{2} V_1 D_1) \circ P_{Q''} - \sqrt{2} D_1 \circ P_{Q'},$$

where $Q' = \{I \in Q : \inf I = \inf I^{(1)}, \ |I| \leq 2^{-1}\}$, $Q'' = \{I \in Q : \sup I = \sup I^{(1)}, |I| \leq 2^{-1}\}$.

Indeed, for $I \in Q'$, we have that

$$W_1(\tilde{\chi}_I) = |\tilde{\chi}_{I+|I|}| - |\tilde{\chi}_I| = -\sqrt{2}\tilde{\chi}_{I^{(1)}} = -\sqrt{2}D_1(\tilde{\chi}_I).$$

For $I \in Q''$, we get that

$$(W_2 + \sqrt{2}V_1 \circ D_1)(\tilde{\chi}_I) = |\tilde{\chi}_{I+2|I|}| - |\tilde{\chi}_I| + \sqrt{2}\tilde{\chi}_{I^{(1)}+|I^{(1)}|} =$$

$$= |\tilde{\chi}_{I+2|I|}| - |\tilde{\chi}_I| + |\tilde{\chi}_{I+|I|}| - |\tilde{\chi}_{I+2|I|}| = W_1(\tilde{\chi}_I).$$

This completes the proof of Theorem 16.c.3. □

Remark 16.c.4. *Let E be an r.i. space with $0 < \alpha_E \le \beta_E < 1$. Then*

$$\|V_l\| \le c(E)\log(2+|l|), \quad \|W_l\|_E \le c(E)\log(2+|l|).$$

This follows from Remark 16.b.4.

Remark 16.c.5. Theorem 16.c.3 gives another proof of Corollaries 11.7 and 11.8.

Lemma 16.c.6. *Let $1 < p < \infty$, $\{h_I\}_{I\in Q} \subset L_p(0,1)$; for every integer l and every $j = 0,1,2,...$*

$$|< h_I, \tilde{\chi}_J >| \le a(l,j)2^{-j/2}, \tag{16.c.7}$$

$$|< h_I, |\tilde{\chi}_{I+l|I|}| >| \le b(l),$$

where $< f,g >= \int_0^1 f(t)g(t)\,dt$; $I,J \in Q$; $J \subset I + l|I| \subset [0,1], |J| = 2^{-j}|I|$; the constants $a(l,j)$ and $b(l)$ do not depend on I and J. Define the operators H and H_1 by the conditions:

$$H(\kappa_{[0,1]}) = H_1(\kappa_{[0,1]}) = 0;$$

$$H(\tilde{\chi}_I) = h_I, \quad H_1(\tilde{\chi}_I) =< h_I, \kappa_{[0,1]} > \kappa_I$$

for $I \in Q$.

If

$$\sum_{l=-\infty}^{\infty}\sum_{j=0}^{\infty} a(l,j)\|V_l\|_{L_p} + \sum_{l=-\infty, l\ne 0}^{\infty} b(l)\|W_l\|_{L_p} < \infty, \tag{16.c.8}$$

then the operator H is bounded in L_p if and only if the operator H_1 is bounded in L_p.

Proof. Let $l \in \mathbf{Z}$, $l \ne 0$. Define the operator $\tilde{W}_l$ by the equations:

$$\tilde{W}_l(\kappa_{[0,1]}) = 0, \quad \tilde{W}_l(\tilde{\chi}_I) =< h_I, |\tilde{\chi}_{I+l|I|}| > W_l(\tilde{\chi}_I)$$

for $I \in Q$. From (16.c.7) and the unconditionality of the H.s. in L_p for $1 < p < \infty$ it follows that

$$\|\tilde{W}_l\|_{L_p} \leq c(p)b(l)\|W_l\|_{L_p}. \tag{16.c.9}$$

Let now $l \in \mathbf{Z}$, $j = 0, 1, 2, \ldots$ Define the operator $\tilde{V}_{l,j}$ by the conditions: $\tilde{V}_{l,j}(\kappa_{[0,1]}) = 0$, $\tilde{V}_{l,j}(\tilde{\chi}_I) = \sum < h_I, \tilde{\chi}_I > \tilde{\chi}_J$ for $I \in Q$, where the sum is taken with respect to $J \in Q$, $J \subset I + l|I|$, $|J| = 2^{-j}|I|$. By Lemma 16.c.1, we have that

$$\|\tilde{V}_{l,j}\|_{L_p} \leq c(p)a(l,j)\|V_l\|_{L_p}. \tag{16.c.10}$$

It remains to observe that

$$(H - H_1)\tilde{\chi}_I = \left(\sum_{l=-\infty, l\neq 0}^{\infty} \tilde{W}_l + \sum_{l=-\infty}^{\infty} \sum_{j=0}^{\infty} \tilde{V}_{l,j} \right)\tilde{\chi}_I \tag{16.c.11}$$

for $I \in Q$. Indeed, we get that

$$(H - H_1)\tilde{\chi}_I = h_I - < h_I, \kappa_{[0,1]} > \kappa_I, \tag{16.c.12}$$

$$\sum_{l=-\infty, l\neq 0}^{\infty} \tilde{W}_l(\tilde{\chi}_I) = \sum_{l=-\infty, l\neq 0}^{\infty} < h_I, |\chi_{I+l|I|}| > W_l(\tilde{\chi}_I) =$$

$$= \sum_{\{l:\ l\neq 0;\ I+l|I|\subset[0,1]\}} < h_I, |\tilde{\chi}_{I+l|I|}| > |\tilde{\chi}_{I+l|I|}| - < h_I, \kappa_{[0,1]\setminus I} > \kappa_I, \tag{16.c.13}$$

$$\sum_{l=-\infty}^{\infty} \sum_{j=0}^{\infty} \tilde{V}_{l,j}(\tilde{\chi}_I) = h_I - S_{|I|^{-1}}(h_I) =$$

$$= h_I - \sum_{\{l: I+l|I|\subset[0,1]\}} < h_I, \tilde{\chi}_{I+l|I|} > |\tilde{\chi}_{I+l|I|}|, \tag{16.c.14}$$

where $S_{|I|^{-1}}$ is the partial Fourier-Haar sum of degree $|I|^{-1}$. Equality (16.c.11) follows from (16.c.12-14); and (16.c.11) and (16.c.8) imply the statement of Lemma 16.c.6. □

16.d. Equivalence of the Haar and the Franklin Systems in L_p, $1 < p < \infty$

We need the following obvious fact.

Lemma 16.d.1. *Let $(n,k) \in \Omega_1$. If the function f is continuous on Δ_n^k and is linear on Δ_{n+1}^{2k-1} and on Δ_{n+1}^{2k}, then*

$$< f, \chi_n^k > = 2^{-n-2} \left(f\left(\frac{k-1}{2^n}\right) - f\left(\frac{k}{2^n}\right) \right).$$

Theorem 16.d.2. *The Haar and the Franklin systems are equivalent in L_p for $1 < p < \infty$.*

Proof. Define the function $\mu : Q \to \{0, 1, ...\}$ in the following way: $\mu(I) = 2^n + k - 1$ when $I \in Q$ is of the form $((k-1)2^{-n}, k2^{-n})$, where $(n, k) \in \Omega_1$. Let $\{f_i\}_0^\infty$ be the Franklin system. Define the operator HF by equations: $HF(\kappa_{[0,1]}) = f_0$, $HF(\tilde{\chi}_I) = f_{\mu(I)}$ for $I \in Q$. This operator maps the H.s. into the Franklin one. Let us prove that HF is bounded in L_p. Observe first that from (16.a.1) and (16.a.2) it follows that, for $I \in Q$,

$$|f_{\mu(I)}| \le c \mid I \mid^{-1/2} \sum_{\{l: I+l|I| \subset [0,1]\}} q^{|l|} \kappa_{I+l|I|}, \tag{16.d.1}$$

where c is an absolute constant, $q = (2+\sqrt{3})^{-1} < 1$. That is why,

$$|< f_{\mu(I)}, | \tilde{\chi}_{I+l|I|} | >| \le cq^l \tag{16.d.2}$$

for $I \in Q$, $I + l|I| \subset [0,1]$. Let $j \ge 1$, $l \in \mathbf{Z}$, $I, J \in Q$, $J \subset I + l|I| \subset [0,1]$, $|J| = 2^{-j}|I|$. By (16.a.1), $f_{\mu(I)}$ is a linear function on $supp\, \tilde{\chi}_J$. By combining Lemma 16.d.1 and inequality (16.d.1), we obtain that

$$\begin{aligned} &|< f_{\mu(I)}, \tilde{\chi}_J >| = 2^{j/2} \mid I \mid^{-1/2} |< f_{\mu(I)}, \tilde{\chi}_J >| \le \\ &\le 2^{j/2} \mid I \mid^{-1/2} 2^{-2} \mid J \mid | f_{\mu(I)}(\inf J) - f_{\mu(I)}(\sup J) | \le \\ &\le 2^{j/2} \mid I \mid^{-1/2} 2^{-2} \mid J \mid 2c \mid I \mid^{-1/2} q^{|l|} 2^{-j} = 2^{-1} c 2^{-3j/2} q^{|l|}. \end{aligned} \tag{16.d.3}$$

Inequalities (16.d.2) and (16.d.3) mean that the sequence $\{f_{\mu(I)}\}_{I \in Q}$ satisfies the assumptions (16.c.7) of Lemma 16.c.6. Furthermore, $< f_{\mu(I)}, \kappa_{[0,1]} >= 0$ for every $I \in Q$. Thus, the continuity of the operator HF in L_p will be proved if we check condition (16.c.8). By Theorem 16.c.3, we have that

$$\begin{aligned} &\sum_{l=-\infty}^{\infty} \sum_{j=0}^{\infty} 2^{-1} c 2^{-j} q^{|l|} \|V_l\|_{L_p} + \sum_{l \in Z, l \neq 0} c q^{|l|} \|W_l\|_{L_p} \le \\ &\le 2^{-1} c\, c(p) \sum_{l=-\infty}^{\infty} \sum_{j=0}^{\infty} 2^{-j} q^{|l|} (\log(2+ \mid l \mid))^{|\frac{1}{2} - \frac{1}{p}|} + \\ &+ c\, c(p) \sum_{l \in Z, l \neq 0} q^{|l|} (\log(2+ \mid l \mid))^{|\frac{1}{2} - \frac{1}{p}|} < \infty. \quad \square \end{aligned}$$

By combining Theorem 16.d.2 and Theorem 1.e.2 we obtain

Corollary 16.d.3. *If E is an r.i. space with non-trivial Boyd indeces, then the Haar and the Franklin systems are equivalent in E.*

Corollary 16.d.3 can be proved directly by using Remarks 16.b.4, 16.c.2 and 16.c.4. Theorem 16.d.2 means that every boundedness result in an r.i. space with non-trivial Boyd indeces for the operator (multiplier or rearrangement) with respect to the H.s. remains true if we replace the H.s. by the Franklin one.

16.e. The Haar and the Franklin systems in r.i. spaces with trivial Boyd indeces

Theorem 16.e.1. *Let E be a separable r.i. space with $\alpha_E = 0$. Then the Haar and the Franklin systems are not equivalent in E.*

Before proving the theorem let us consider, for positive integers n, the functions

$$q^{(n)}(t) := \sum_{i=2}^{n} 2^{-i/2}\tilde{\chi}_{2^i+3}(t), \quad t \in [0,1].$$

Obviously, $\|q^{(n)}\|_{L_\infty} = 1$. Denote

$$f^{(n)}(t) = \sum_{i=2}^{n} 2^{-i/2} f_{2^i+2}(t), \quad t \in [0,1].$$

From (16.a.2) it follows that

$$f^{(n)}(0) = \sum_{i=2}^{n} 2^{-i/2} f_{2^i+3}(0) \le \sum_{i=2}^{n} 2^{-i/2}(-C_1)2^{i/2}q^3 = (1-n)C_1 q^3.$$

Thus, the operator HF, defined by equations $HF\tilde{\chi}_m = f_{m-1}$ (the tilde above χ means that the Haar function is normalized in L_2), is unbounded in L_∞ on the polynomials with respect to the H.s. The idea of the proof of Theorem 16.e.1 consists in the transference of this example to an arbitrary r.i. space E with $\alpha_E = 0$ using Theorem 1.d.3.

Proof of Theorem 16.e.1. Fix the integer $n \ge 13$. Put $N := 2^n$, $n_0 := n - 11$. Since $\alpha_E = 0$, it follows from Theorem 1.d.3 that there exists a sequence $\{x_i\}_{i=1}^N$ of equimeasurable pairwise disjoint normalized functions which is 2-equivalent to the unit vector basis of l_∞^N. As it was explained in Theorem 6.3 we can assume that the functions x_i have the following form:

$$x_i = \sum_{j=1}^{m} a_j \kappa_{((i-1+(j-1)N)2^{-k},(i+(j-1)N)2^{-k})}, \tag{16.e.1}$$

where the scalars a_j and the positive integers m and k do not depend on i. For convenience we suppose that

$$a_1 > a_2 > \ldots > a_m > 0. \tag{16.e.2}$$

Put

$$g(t) = \sum_{i=2}^{n_0} 2^{-i/2} \tilde{\chi}_{2^i+3}(t), t \in [0,1],$$

$$h(t) = \sum_{j=1}^{m} a_j g_j(t), \quad t \in [0,1],$$

where

$$g_j(t) = \begin{cases} g(2^{k-n}t - j + 1), & t \in ((j-1)2^{n-k}, j2^{n-k}); \\ 0, & t \notin ((j-1)2^{n-k}, j2^{n-k}). \end{cases}$$

From (16.e.1) it follows that the function h is a linear combination of $\{x_i\}_{i=1}^N$ with coefficients equal to ± 1. That is why

$$\|h\|_E \leq 2. \tag{16.e.3}$$

It is clear that

$$g_j = \sum_{i=2}^{n_0} 2^{-\frac{k+i-n}{2}} \tilde{\chi}_{k+i-n}^{(j-1)2^i+3}$$

for $j = 1, ..., m$. Put

$$\mu(j,i) = 2^{k+i-n} + (j-1)2^i + 2,$$

$$\nu(j,i) = (j-1)2^i + 2,$$

$$f(t) := (HFh)(t) = \sum_{j=1}^{m} a_j \sum_{i=2}^{n_0} 2^{-\frac{k+i-n}{2}} f_{\mu(j,i)}(t),$$

where $j = 1, ..., m$; $i = 2, ..., n_0$, and $t \in [0,1]$.

We are going to prove that

$$\|f\|_E \geq (n_0 - 1)\left(\frac{c_1 q^3}{2} - \frac{c_2}{(2+\sqrt{3})^8 - 1}\right), \tag{16.e.4}$$

where c_1, c_2, q are constants from (16.a.2).

Let $\tau_j^{(1)} := (j-1)2^{n-k}$, $\tau_j^{(2)} := (j-1)2^{n-k} + 2^{-k}$, $I_j := (\tau_j^{(1)}, \tau_j^{(2)})$ for $j = 1, ..., m$. From (16.a.1) it follows that $\tau_j^{(1)} = t_{2\nu(j,i)-4}^{(\mu(j,i))}$ for $j = 1, ..., m$ and $i = 2, ..., n_0$. That is why (16.a.2) implies that

$$f_{\mu(j,i)}(\tau_j^{(1)}) \leq -2^{\frac{k+i-n}{2}} c_1 q^3. \tag{16.e.5}$$

Since the function $f_{\mu(j,i)}$ is a linear one on I_j, we have that

$$f_{\mu(j,i)}(\tau_j^{(2)}) = (1 - 2^{i+1-n}) f_{\mu(j,i)}(\tau_j^{(1)}) + 2^{i+1-n} f_{\mu(j,i)}(t_{2\nu(j,i)-3}^{(\mu(j,i))}).$$

By (16.a.2),

$$f_{\mu(j,i)}(\tau_j^{(2)}) \leq -2^{\frac{k+i-n}{2}}(c_1q^3(1-2^{i+1-n}) - c_2q^2 2^{i+1-n}).$$

If $i \leq n_0 = n - 11$, then

$$c_1q^3(1-2^{i+1-n}) - c_2q^2 2^{i+1-n} \geq \frac{1}{2}c_1q^3.$$

Finally, we get that

$$f_{\mu(j,i)}(\tau_j^{(2)}) \leq -\frac{c_1q^3}{2}2^{\frac{k+i-n}{2}} \tag{16.e.6}$$

for $j = 1, \ldots, m$; $i = 2, \ldots, n_0$.

In what follows we shall use the folowing properties of the Franklin system proved in [64, 65]. Let

$$\zeta = 2^\xi + \eta, \quad \xi, \eta \in \mathbf{Z}; \quad \xi \geq 0, \quad 1 < \eta < 2^\xi.$$

Then

$$f_\zeta(t_{2\eta-1}^{(\zeta)}) > -f_\zeta(t_{2\eta-2}^{(\zeta)}) > \ldots > f_\zeta(t_1^{(\zeta)}) > -f_\zeta(t_0^{(\zeta)}), \tag{16.e.7}$$

$$f_\zeta(t_{2\eta-1}^{(\zeta)}) > -f_\zeta(t_{2\eta}^{(\zeta)}) > \ldots > (-1)^{(\zeta)}f_\zeta(t_{\zeta-1}^{(\zeta)}) > (-1)^{\zeta+1}f_\zeta(t_\zeta^{(\zeta)}). \tag{16.e.8}$$

Consider $r \in \{j+1, \ldots, m\}$, where $j = 1, \ldots, m-1$. It is clear that $\tau_j^{(2)} < t_{2(j-1)2^i+1}^{(\mu(r,i))}$ for $i = 2, \ldots, n_0$. From (16.e.7), (16.e.2), (16.a.2) it follows that, for $t \in I_j$,

$$\begin{aligned} &|\textstyle\sum_{r=j+1}^m a_r \sum_{i=2}^{n_0} 2^{-\frac{k+i-n}{2}} f_{\mu(r,i)}(t)| \leq \\ &\leq a_j \textstyle\sum_{r=j+1}^m \sum_{i=2}^{n_0} c_2 q^{(r-j)2^{i+1}+2} \leq \\ &\leq c_2 a_j(n_0-1)\textstyle\sum_{r=j+1}^m (q^8)^{r-j} \leq \\ &\leq c_2(n_0-1)a_j\frac{q^8}{1-q^8} = \frac{c_2a_j(n_0-1)}{(2+\sqrt{3})^8-1}. \end{aligned} \tag{16.e.9}$$

Let now $r \in \{1, \ldots, j-1\}$, where $j = 2, \ldots, m$. In this case $\tau_j^{(1)} = t_{\nu_1(j,r,i)}^{\mu(r,i))}$, where $\nu_1(j,r,i) = (j+r-2)2^i + 2$, $\quad i = 2, \ldots, n_0$. From (16.a.2) it follows that

$$f_{\mu(r,i)}(\tau_j^{(1)}) < 0 \tag{16.e.10}$$

for $r = 1, \ldots, j-1$; $j = 2, \ldots, m$. Besides, $t_{\nu_1(j,r,i)+1}^{(\mu(r,i))} - t_{\nu_1(j,r,i)}^{(\mu(r,i))} > 2(\tau_j^{(2)} - \tau_j^{(1)})$. That is why (16.e.8) and (16.e.10) imply that

$$f_{\mu(r,i)}(t) < 0 \tag{16.e.11}$$

for $t \in I_j$, $j = 2, ..., m$; $r = 1, ..., j-1$; $i = 2, ..., n_0$.

We are prepared now to estimate $f(t)$ for $t \in I_j$, $j = 1, ..., m$. Indeed, it follows from (16.e.5-6), (16.e.9), (16.e.11) that

$$f(t) = \sum_{r=1}^{j-1} a_r \sum_{i=2}^{n_0} 2^{-\frac{k+i-n}{2}} f_{\mu(r,i)}(t) + a_j \sum_{i=2}^{n_0} 2^{-\frac{k+i-n}{2}} f_{\mu(j,i)}(t) +$$

$$+ \sum_{r=j+1}^{m} a_r \sum_{i=2}^{n_0} 2^{-\frac{k+i-n}{2}} f_{\mu(r,i)(t)} \leq$$

$$\leq -(n_0 - 1)\frac{c_1 a_j q^3}{2} + (n_0 - 1)\frac{c_2 a_j}{(2+\sqrt{3})^8 - 1}.$$

Taking into account (16.e.1), we get that

$$| f(t) | \geq (n_0 - 1)\left(\frac{c_1 q^3}{2} - \frac{c_2}{(2+\sqrt{3})^8 - 1}\right) x_1(t)$$

for $t \in [0,1]$. Since $\|x_1\|_E = 1$, the last inequality implies (16.e.4). The inequalities (16.e.3) and (16.e.4) mean that the operator HF is unbounded in E, since n is an arbitrary positive integer. □

Corollary 16.e.2. *The Haar and the Franklin systems are not equivalent in a separable r.i. space E with $\beta_E = 1$.*

Proof. Consider the operator FH defined by the equations: $FHf_i = \tilde{\chi}_{i+1}$ for $i = 0, 1, 2, ...$ It is clear that the operator HF is adjoint to FH. Since $\alpha_{E'} = 1 - \beta_E = 0$ (see (1.d.5)), it follows from Theorem 16.e.1 that the operator HF is unbounded in E'. Consequently, the operator FH is unbounded in E. □

Remark 16.e.3. *Theorem 16.a.1 can be strengthened. The operator HF is bounded in a separable r.i. space E if and only if $0 < \alpha_E \leq \beta_E < 1$. The same is valid for the operator FH by the duality argument.*

To prove these assertions it suffices to check that the boundedness of the operator FH in E imply that $\alpha_E > 0$. The proof of the last statement is similar to that of Theorem 16.e.1. However, the calculations are by far more complicated. That is why we give only the scheme of reasoning.

Let $\alpha_E = 0$. Fix $n \in \mathbf{N}$. Consider the sequence $\{x_i\}_{i=1}^N$ for $N = 2^{2n+7}$, which is the same as in the proof of Theorem 16.e.1. Put

$$u(t) = \sum_{i=4}^{n} \chi_{2^i - 2}(t),$$

$$w(t) = \sum_{j=1}^{m} a_j u_j(t), \quad t \in [0,1],$$

where

$$u_j(t) = \begin{cases} u(2^{k-n}t - (j-1)2^{n+7}) \\ \quad \text{for } t \in ((j-1)N2^{-k}, (j-1)N2^{-k} + 2^{n-k}); \\ \\ 0, \quad \text{otherwise.} \end{cases}$$

From (16.e.1) it follows that w is a linear combination of $\{x_i\}_1^N$ with coefficients equal to ± 1. That is why

$$\|w\|_E \leq 2. \tag{16.e.12}$$

However,

$$\| \sum_{r=2^{k-n+2}}^{2^{k-1}-1} \langle f_r, w \rangle \tilde{x}_{r+1} \|_E \leq c(n-3), \tag{16.e.13}$$

where c is a constant independent on n; $\langle x, y \rangle := \int_0^1 x(t)y(t)\,dt$. It is clear that inequalities (16.e.12), (16.e.13) mean that the operator FH is unbounded in E since n is an arbitrary positive integer.

The proof of (16.e.13) is based on three auxiliary lemmas. We shall give them without proofs. Let $j = 1, ..., m$; $i = 4, ..., n$; $\nu(j,i) := (j-1)2^{n+i+5} + 2^{i-2} - 1$, $\mu(j,i) := 2^{k+i-n-2} + \nu(j,i)$.

Lemma 16.e.4. *For every $j = 1, ..., m$ and every $i = 4, ..., n$ we have that*

$$\langle f_{\mu(j,i)}, u_j \rangle \quad \geq \quad 2^{-k-i+n-2} f_{\mu(j,i)}(t^{(\mu(j,i))}_{2\nu(j,i)-1}). \tag{16.e.14}$$

Lemma 16.e.5. *For every $j = 1, ..., m$ and every $i = 4, ..., n$ we get that*

$$\sum_{r=j+1}^{m} |\langle f_{\mu(j,i)}, u_r \rangle| \quad < \quad \frac{1}{3} 2^{-k+n-i-2} f_{\mu(j,i)}(t^{(\mu(j,i))}_{2\nu(j,i)-1}). \tag{16.e.15}$$

Lemma 16.e.6. *For every $j = 2, ..., m$ and every $i = 1, ..., j-1$,*

$$\langle f_{\mu(j,i)}, u_r \rangle \quad \geq \quad 0. \tag{16.e.16}$$

From (16.e.2), (16.e.14-16), (16.a.2) it follows that

$$\begin{aligned} &\langle f_{\mu(j,i)}, w \rangle = \textstyle\sum_{r=1}^{j-1} \langle f_{\mu(j,i)}, u_r \rangle a_r + \\ &+ a_j \langle f_{\mu(j,i)}, u_j \rangle + \textstyle\sum_{r=j+1}^{m} a_r \langle f_{\mu(j,i)}, u_r \rangle \quad > \\ &> \quad a_j 2^{-k-i+n-2} f_{\mu(j,i)}(t^{(\mu(j,i))}_{2\nu(j,i)-1}) - \\ &- a_j \textstyle\sum_{r=j+1}^{m} |\langle f_{\mu(j,i)}, u_r \rangle| \quad \geq \\ &\geq \tfrac{2}{3} a_j 2^{-k-i+n-2} f_{\mu(j,i)}(t^{(\mu(j,i))}_{2\nu(j,i)-1}) \geq \tfrac{2}{3} c_1 a_j 2^{-\frac{k+i-n}{2}}. \end{aligned} \tag{16.e.17}$$

If $t \in ((j-1)N2^{-k} + 2^{n-k} - 2^{-k+2}, (j-1)N2^{-k} + 2^{n-k})$ for $j = 1, \ldots, m$, then, by (16.e.17), we have that

$$| \sum_{r=2^{k-n+2}}^{2^{k-1}-1} \langle f_r, w\rangle \tilde{\chi}_{r+1}(t) | = | \sum_{i=4}^{n} \langle f_{\mu(j,i)}, w\rangle \tilde{\chi}_{\mu(j,i)+1}(t) | \geq$$

$$\geq \tfrac{2}{3} c_1 a_j \sum_{i=4}^{n} 2^{-\frac{k+i-n}{2}} 2^{\frac{k+i-n}{2}} = \tfrac{2}{3} c_1 a_j (n-3).$$

The last inequality implies that

$$| \sum_{r=2^{k-n+2}}^{2^{k-1}} \langle f_r, w\rangle \tilde{\chi}_{r+1} | \geq \frac{2}{3} c_1 (n-3) \sum_{p=2^n-3}^{2^n} x_p.$$

Thus, (16.e.13) is proved. □

Chapter 17

Olevskii System

A.M. Olevskii investigated some orthonormal system which is closely connected with the H.s.[212].

At first we construct a sequence of orthogonal matrices $A_m = (a_{ij}^{(m)})$, $m = 1, 2, \ldots$, $1 \le i, j \le 2^m$. Put $a_{1j} = 2^{-\frac{m}{2}}$, $1 \le j \le 2^m$. If $i = 2^r + s$, $0 \le r \le m-1$, $1 \le s \le 2^r$, then we put

$$a_{ij}^m = \begin{cases} 2^{-\frac{m-r}{2}}, & (s-1)2^{m-r} < j \le (2s-1)2^{m-r-1} \\ -2^{-\frac{m-r}{2}}, & (2s-1)2^{m-r-1} < j \le s2^{m-r} \\ 0, & \text{for other } i, j. \end{cases}$$

Denote by $\tilde{\chi}_n(t)$ or $\tilde{\chi}_n^k(t)$ the Haar functions normed in L_2. Put $\varphi_1 = \chi_1$, $\varphi_2 = \chi_2$,

$$\varphi_n = \sum_{i=1}^{2^m} a_{ij}^{(m)} \tilde{\chi}_m^i,$$

where $n = 2^m + j$, $m = 1, 2, \ldots$, $1 \le j \le 2^m$. The system $\{\varphi_n\}$ is orthonormal. It forms a basis in L_1 and C [212]. Using this statement and Theorem 1.e.3 we can prove that $\{\varphi_n\}$ is a basis in any separable r.i. space. K.S.Kazarian proved that it does not form an unconditional basis in L_p $(p \ne 2)$ [138]. This statement may be strengthened.

Theorem 17.1. *Let the Olevskii system $\{\varphi_n\}$ be an unconditional basis in a separable r.i. space E. Then $E = L_2$ up to equivalence.*

Proof. Without loss of generality we may assume that $\|\mathbf{1}\|_E = 1$. The definition of the matrices A_m implies that

$$\chi_k^1 = 2^{-k} \sum_{j=1}^{2^k} \varphi_{2^k+j} \tag{17.1}$$

for every $k = 1, 2, \ldots$. By Theorem 1.b.8 there exists a constant $C > 0$ such that

$$\frac{1}{C}\|(\sum_k (c_k\varphi_k)^2)^{\frac{1}{2}}\|_E \le \|\sum_k c_k\varphi_k\|_E \le C\|(\sum_k (c_k\varphi_k)^2)^{\frac{1}{2}}\|_E \tag{17.2}$$

for every sequence of reals $\{c_k\}$.

Given sequences $\{c_k\}$ and $\{\varepsilon_k = \pm 1\}$, we consider the function

$$f_\varepsilon(t) = \sum_{k=1}^{\infty} \varepsilon_k c_k \chi_k^1(t).$$

Using (17.1), we get

$$f_\varepsilon = \sum_{k=1}^{\infty} \varepsilon_k c_k 2^{-k} \sum_{j=1}^{2^k} \varphi_{2^k+j}.$$

By (17.2)

$$\|f_\varepsilon\|_E \le C\|(\sum_{k=1}^{\infty}(c_k 2^{-k})^2)^2 \sum_{j=1}^{2^k}(\varphi_{2^k+j})^2)^{\frac{1}{2}}\|_E.$$

Since

$$\sum_{j=1}^{2^k} \varphi_{2^k+j}^2(t) = \sum_{j=1}^{2^k}\left(\sum_{i=1}^{2^k} a_{ij}^{(k)} \tilde{\chi}_k^i(t)\right)^2 =$$

$$= 2^k \sum_{j=1}^{2^k}\sum_{i=1}^{2^k}(a_{ij}^{(k)})^2 \kappa_{(\frac{i-1}{2^k}, \frac{i}{2^k})}(t) =$$

$$= 2^k \sum_{i=1}^{2^k}\sum_{j=1}^{2^k}(a_{ij}^{(k)})^2 \kappa_{(\frac{i-1}{2^k}, \frac{i}{2^k})}(t) = 2^k,$$

then

$$\|f_\varepsilon\|_E \le C\|(\sum_{k=1}^{\infty}(c_k 2^{-k})^2 2^k)^{\frac{1}{2}}\|_E = C(\sum_{k=1}^{\infty} c_k^2 2^{-k})^{\frac{1}{2}}.$$

On the other hand,

$$\sup_{\varepsilon_k=\pm 1} \|\sum_{k=1}^{\infty}\varepsilon_k c_k \chi_k^1\|_E \ge \frac{1}{2}\|\sum_{k=1}^{\infty} | c_{2k} | \chi_{2k}^1\|_E + \frac{1}{2}\|\sum_{k=1}^{\infty} | c_{2k-1} | \chi_{2k-1}^1\|_E.$$

If $t \in (2^{-2k-2}, 2^{-2k-1})$, then

$$\sum_{n=1}^{\infty}|c_{2n}|\chi_{2n}^1(t) = \sum_{n=1}^{k}|c_{2n}| \ge |c_{2k}|.$$

If $t \in (2^{-2k-1}, 2^{-2k})$, then

$$\sum_{n=1}^{\infty}|c_{2n-1}|\chi_{2n-1}^1(t) = \sum_{n=1}^{k}|c_{2n-1}| \ge |c_{2k-1}|.$$

Therefore

$$\sup_{\varepsilon_k=\pm 1} \| \sum_{k=1}^{\infty} \varepsilon_k c_k \chi_k^1 \|_E \geq \tfrac{1}{2} \| \sum_{k=1}^{\infty} | c_{2k} | \kappa_{(2^{-2k-2},2^{-2k-1})} \|_E +$$

$$+ \tfrac{1}{2} \| \sum_{k=1}^{\infty} | c_{2k-1} | \kappa_{(2^{-2k-1},2^{-2k})} \|_E \geq \tfrac{1}{2} \| \sum_{k=1}^{\infty} | c_k | \kappa_{(2^{-k-2},2^{-k-1})} \|_E.$$

So,

$$\| \sum_{k=1}^{\infty} | c_k | \kappa_{(2^{-k-2},2^{-k-1})} \|_E \leq C (\sum_{k=1}^{\infty} c_k^2 2^{-k})^{\frac{1}{2}} = \\ = 2\sqrt{2} C \| \sum_{k=1}^{\infty} c_k \kappa_{(2^{-k-1},2^{-k})} \|_{L_2}. \tag{17.3}$$

Given $x = x^* \in E$, we construct the function

$$y(t) = \sum_{k=1}^{\infty} x(2^{-k}) \kappa_{(2^{-k-1},2^{-k})}(t).$$

It is evident that, $0 \leq y(t) \leq x(t) \leq y(\frac{t}{2})$. The operator $\sigma_2 x(t) = x(\frac{t}{2})$ is bounded in E and $\|\sigma\|_E \leq 2$ (Section 1.d). Therefore

$$\|y\|_E \leq \|x\|_E \leq 2\|y\|_E.$$

Applying (17.3), we get

$$\|x\|_E \leq 8\sqrt{2} C \|x\|_{L_2} \tag{17.4}$$

for every $x = x^* \in E$. Clearly, (17.4) is valid for every $x \in L_2$.

If the Olevskii system is an unconditional basis in E, then it is an unconditional basis in $(E')_0$ too (Section 1.d) with the same basis constant. Consequently,

$$\|x\|_{E'} \leq 8\sqrt{2} C \|x\|_{L_2}$$

for every $x \in L_2$. Hence,

$$\frac{1}{8\sqrt{2}C} \|x\|_{L_2} \leq \|x\|_{E''}$$

for every $x \in E''$. The embedding $E \subset E''$ is isometric. Therefore, for each $x \in E$,

$$\frac{1}{8\sqrt{2}C} \|x\|_{L_2} \leq \|x\|_E.$$

The obtained inequality and (17.4) prove the theorem. □

So, the Olevskii system is a conditional basis in any separable r.i. space $E \neq L_2$.

References

[1] Abdulgamidov A.R. Orthonormal Haar systems defined on bounded closed sets of R^n, *Uchenye zapiski Kazan Univ.*, 128(5):3–19, 1968. (in Russian)

[2] Abdulgamidov A.R. On approximation of respect to the H.s., *Uchenye zapiski Kazan Univ.*, 129(3):30–52, 1969. (in Russian)

[3] Agayan S.S. On stability of summation of Fouirier-Haar series with approximate coefficients. *Mat. zametki*, 39(1):136–143, 1986. (in Russian)

[4] Agayan S.S., Metevosyan A.K., and Muradyan A.L. Digital filters with respect to the family of the Haar systems. *Dokl. Acad. Nauk Arm. SSR*, 77:117–121, 1983. (in Russian)

[5] Aidosov E.G. On relations between modulus of continuity and the best approximations with respect to the H.s. *Izv. Acad. Nauk*, 138(5):3–7, 1987. (in Russian)

[6] Ando T. Contractive projections in L_p spaces. *Pacific j. Math.*, 17:391–405, 1966.

[7] Andrew A.D. A remark on generalized Haar system in L_p, $1 < p < \infty$. *Pacific j. Math.*, 82(2):317–321, 1979.

[8] Andrew A.D. On subsequences of the Haar system in $C(\Delta)$. *Isr. j. Math.*, 31(1):85–90, 1978.

[9] Andrew A.D. Pertubations of Schauder Basis. *Studia Math.*, 65(3):287–298, 1979.

[10] Antipa A. Doob's Inequality for Rearrangement Invariant Function Spaces. *Rev. Roumaine Math. Pures Appl.*, 35(2):101–108, 1990.

[11] Arutyunyan F.G. On the uniqueness of Haar series. *Dokl. Acad. Nauk Arm. SSR*, 38(3):129–134, 1964. (in Russian)

[12] Arutyunyan F.G. On the Haar series. *Dokl. Acad. Nauk Arm. SSR*, 43(3):134–140, 1966. (in Russian)

[13] Arutyunyan F.G. Reconstruction of coefficients of the Haar and of the Walsh series converging to functions which are Denjoy integrable. *Izv. Acad. Nauk SSSR Ser. Mat.*, 30(2):324–344, 1966. (in Russian)

[14] Arutyunyan F.G. and Talalyan A.A. On the uniqueness of the Haar and of the Walsh series. *Izv. Acad. Nauk SSSR Ser. Mat.*, 28(6):1391–1408, 1964. (in Russian)

[15] Assani I. Sur les operateurs a puissances bornees et le theoreme ergodicue pontuel dans $L_p[0,1]$, $1 < p < \infty$. *Can. j. Math.*, 38(7):937–946, 1986.

[16] Astashkin S.V. The example of R.I. space of non-fundamental type. *XIV school on the theory of operators in function spaces. Novgorod*, p.5, 1989. (in Russian)

[17] Balashov L.A. On the Haar series. *Mat. zapiski*, 10(4):369–374, 1971. (in Russian)

[18] Bareladze G.P. On absolute and unconditional convergence of multiple Haar series. *Soobscheniya Acad. Nauk GSSR*, 111(1):21–24, 1983. (in Russian)

[19] Bari N.K. *Trigonometric series.* Moscow: Fizmatgiz, 1961. (in Russian)

[20] Beausamy B. Introduction to Banach spaces and their geometry. *Math. Studies*, 68:1–417, 1985.

[21] Bennet C. and Sharley R. *Interpolation of Operators.* London: Academic Press, 1988.

[22] Bollobas B. Martingale inequalities. *Math. Proc. Cambridge Philos. Soc.*, 87:377–382, 1980.

[23] Bergh J. and Lofstrom J. *Interpolation Spaces: An Introduction.* Springer, 1976.

[24] Bochkarov S.V. On unconditional almost every where convergence of Fourier-Haar series of continuous functions. *Mat. zametki*, 4(2):211–220, 1968. (in Russian)

[25] Bochkarov S.V. On the Haar series. *Uspehi Mat. Nauk*, 24(1):189–190, 1969. (in Russian)

[26] Bochkarov S.V. On the coefficients of series with respect to the Haar system. *Mat. Sb.*, 80(1):97–116, 1969. (in Russian)

[27] Bochkarov S.V. The method of averaging in the theory of orthonormal series and some questions of the theory of functions. *Trudy MIAM*, 146:3–87, 1978. (in Russian)

[28] Bokaev N.L. On the Fourier-Haar coefficients. *Izv. Acad. Nauk*, (3):13–16, 1990. (in Russian)

[29] Bourgain J. A counterexample to a complementation problem. *Compositio Math.*, 43(1):133–144, 1981.

[30] Boyanov B.D. and Veselinov V.M. On the approximation of functions by Haar series. *Matematica (RSR)*, 14(2):189–192, 1972.

[31] Boyd D.W. Indices functional spaces and their relationship to interpolation. *Canadian j. Math.*, 21:1245–1254, 1969.

[32] Bru B. and Heinich H. Applications de dualite dans les espaces de Kothe. *Studia Math.*, 93:41–69, 1989.

[33] Bru B. and Heinich H. Derivations et constractions dans espaces de Kothe. *C.R. Acad. Sci.*, 301(4):131–134, 1985.

[34] Brunetti M.E. and Zanell V. Matrici di intergazione frazionaria per il sistema di Haar. *Atti Semin mat. e fis Univ. Modena*, 38(1):179–208, 1990.

[35] Bryil'kov A.N. On the sets of convergence of the Haar series to infinity. *Technical Report 7287-B87, Dep. VINITI, 15.10.87.* (in Russian)

[36] Bryil'kov A.N. On the sets of points of unbounded divergence of the Haar series. *Vestnik MGU Mat., mech.*, 2:86–88, 1988. (in Russian)

[37] Bugadze V.M. On the Fourier-Haar coefficients. *Soobscheniya Acad. Nauk GSSR*, 103:37–39, 1981. (in Russian)

[38] Bugadze V.M. *Fourier-Haar series of superposition of functions.* Tbilisi, 1988. (in Russian)

[39] Bugadze V.M. On the changes of variables preserving absolute convergence of the Fourier-Haar coefficients. *Soobscheniya Acad. Nauk GSSR*, 134(3):486–487, 1989. (in Russian)

[40] Bugadze V.M. On the Fourier-Haar series of superpositions. *Mat. Sb.*, 182(2):175–202, 1991. (in Russian)

[41] Burkholder D.L. A nonlinear partial differential equation and the unconditional constant of the Haar system in L_p. *Bull. Amer. Math. Soc., New Series*, 7:591–595, 1982.

[42] Burkholder D.L. A nonlinear partial differential equation and the unconditional constant of the Haar system in L_p. *Bull. Amer. Math. Soc., New Series*, 7:591–595, 1982.

[43] Burkholder D.L. A proof of Pelczynski's conjecture for the Haar system. *Stud. Math.*, 91(1):79–83, 1988.

[44] Burkholder D.L. A sharp inequality for martingale transforms. *Ann. Probability*, 7:858–863, 1979.

[45] Burkholder D.L. An elementary proof of an inequality of R.E.A.C. Paley. *Bull. Lond. Math. Soc.*, 17:474–478, 1985.

[46] Burkholder D.L. Boundary value problems and sharp inequalities for martingale transforms. *Ann. Probab.*, 12:647–702, 1984.

[47] Burkholder D.L. Distribution function inequalities for martingales. *Proc., Sixth Berkeley Symp. Math. Statist. Prob.*, 2:223–240, 1972.

[48] Burkholder D.L., Davis B.J., and Gundy R.F. Integral inequalities for convex functions of operators on martingales. volume 2, pages 223–240. Proc. Sixth Berkeley Symp. Math. Statist. Prob., 1972.

[49] Burkholder D.L. and Gundy R. Extrapolation and interpolation of quasi - linear operators of martingales. *Acta Math.*, 124:249–303, 1970.

[50] Calderon A.P. Intermediate spaces and interpolation, the complex method. *Studia Math.*, 24:113–190, 1964.

[51] Calderon A.P. Spaces between L_1 and L_∞ and the theorem of Marcinkiewicz. *Studia Math.*, 26:273–299, 1966.

[52] Carleson L. An explicit unconditional basis in H^1. *Bull. Sci. Math.*, 104:405–416, 1980.

[53] Carothers N.L., Dilworth S.J., and Trautman D.A. On the geometry of the unit sphere of the Lorentz space $L_{W,1}$. 34:21–27, 1992.

[54] Chanturija Z.A. Absolute convergence of the Fourier-Haar coefficients. *Soobscheniya Acad. Nauk GSSR*, 75(4):281–284, 1964. (in Russian)

[55] Chanturija Z.A. On absolute convergence of the series of Haar-Fourier coefficients. volume 2, pages 25–35. Comment. Math. Prace Mat., special issue (dedicated to Wladyslaw Orlicz on the occassion of his seventy-fifth birthday), 1979.

[56] Chen Guoting. Best rational approximation with denominators having restricted ranges. *Technical report, Math. Res. Rep.*, Prepr. 19, 1985.

[57] Chen Jan-D. and Chu Liang-Ju. On convergence of Haar series to infinity. *Rocz. PTH. ser. 1*, 27(1):39–48, 1987.

[58] Chgan Minlyan. On convergence of Haar series. *Proc. Conf. on approximation theory, Khanchgon*, p. 136–137, 1978.

[59] Chkheidze G.A. On summation of multiple Haar series. *Some questions of theory of functions, Tbilisi*, 2:128–163, 1981. (in Russian)

[60] Ciesielski Z. Equivalence and shift property of spline bases in L_p spaces. *Constructive function theory. Proc. Int. Conf. Blagoevgrad, 1977*, p. 281–291, 1980.

[61] Ciesielski Z. Haar systems and nonparametric deneity estimation in several variable. *Prob. Math. Stat.*, 9(2):1–11, 1988.

[62] Ciesielski Z. On Haar functions and on the Schauder basis of the space $C(0,1)$. *Bull. Acad. Polon. Sci.*, 7:227–232, 1959.

[63] Ciesielski Z. Properties of the orthonormal Franklin system I, II. *Studia Math.*, 27(3):141–157, 1963.

[64] Ciesielski Z. Some properties of the Schauder bases of the space $C(0,1)$. *Bull. Acad. Polon. Sci.*, (8):141–144, 1960.

[65] Ciesielski Z. and Figiel T. Spline bases in classical function spaces on compact C^∞ manifolds, II. *Studia Math.*, 76:95–136, 1983.

[66] Ciesielski Z. and Kwapien S. Some properties of the Haar, Walsh-Paley, Franklin and the bounded polygonal orthogonal bases in L_p spaces. *Comment. Math. Prace Mat., special issue (dedicated to Wladislaw Orlicz on the occassion of his seventyfifth birthday)*, 2:37–42, 1979.

[67] Ciesielski Z. and Musielak J. On absolute convergence of Haar series. *Colloq. math.*, 7(1):61–65, 1959.

[68] Ciesielski Z., Simon P., and Sjolin P. Equivalence of Haar and Franklin bases in L_p spaces. *Studia Math.*, 60(2):195–211, 1979.

[69] Cox D.C. The best constant in Burkholder's weak - L^1 inequality for the martingale square function. *Proc. Amer. Math. Soc.*, 85:427–433, 1981.

[70] Davis B. On the integrability of the martingale square function. *Isr. J. Math.*, 8:187–190, 1970.

[71] De Leew K., Katztnelson Y., and Kahane J.P. Sur les coefficients de Fouirer des fonctions continues. *C. R. Acad. Sci.*, 285(16):1001–1003, 1977.

[72] Delporte J. Representation des fonctions continues de deux variables et des covariances continues par des developpements en series de Haar uniformement convergents. *C.R. Acad. Sci.*, 260(10):2701–2704, 1965.

[73] Dilworth S.J. Some probabilistic inequalities with application to functional analysis. 144:53–67, 1993.

[74] Dilworth S.J. Intersection of Lebesque spaces L_1 and L_2. *Proc. Amer. Math. Soc.*, 103(4):1185–1188, 1988.

[75] Doob J.L. *Stochastic processes.* N.-Y.: John Waley, 1953.

[76] Dmitriev A.A. and Semenov E.M. On the operators of weak type $(1,1)$. *Sib. Mat. J.*, 20(3):656–658, 1979. (in Russian)

[77] Dor L.E. On projections in L_1. *Ann. Math.*, 102:463–474, 1975.

[78] Dor L. and Odell E. Monotone bases in L_p. *Pacif. J. Math.*, 60(2):5–61, 1976.

[79] Doroshenko O.N. and Musaev M.M. Power polinomial approximation with the help of multiplicative Haar bases. *Voprosy kibernetiki, Tashkent*, 106:140–150, 1977. (in Russian)

[80] Douglas R.G. Contractive projections on an L_1 spaces. *Pacif. J. Math.*, 15:443–462, 1965.

[81] Ebralidze A.D. On Haar and Rademacher series. *Soobsecheniya Acad. Nauk GSSR*, 83(2):297–300, 1976. (in Russian)

[82] Ebralidze A.D. On uniqueness of multiple Haar series. *Soobscheniya Acad. Nauk*, 70(3):537–539, 1973. (in Russian)

[83] Ebralidze A.D. On Fourier-Haar coefficients of continuously differentiable functions. *Soobscheniya Acad. Nauk*, 94(1):29–31, 1979. (in Russian)

[84] Efimov A.V. Linear methods of approximation of continuous periodic functions. *Mat.Sb.*, 54(1):51–90, 1961. (in Russian)

[85] Ellis H.W. and Halperin I. Haar fucntions and the basis problem for Banach spaces. *J. London Math. Soc.*, 31:28–39, 1956.

[86] Enflo P. and Starbird T. Subspaces of L_1 containing L_1. *Studia Math.*, 65:203–224, 1979.

[87] Faber G. Uber die Orthogonalfunktionen des Haar. *Jahresber. Deutsch. Math. Verein*, 9:104–112, 1910.

[88] Fefferman C. and Stein E. Some maximal inequality. *Amer. J. Math.*, 93:107–115, 1971.

[89] Figiel T. On equivalence of some bases to the Haar system in spaces of vector-valued fucntions. *Bull. Pol. Acad. Sci. Math.*, 36(3-4):119–131, 1988.

[90] Fine N.J. On Walsh functions. *Trans. Amer. math. Soc.*, 65:372–414, 1949.

[91] Franchetti C. The norm of the minimal projection onto hyperplanes in $L^p[0,1]$ and the radial constant. *Boll. U.M.I.*, 7(4-B):803–821, 1990.

[92] Franklin Ph. A set of continuous orthogonal functions. *Math. Ann.*, 100:522–529, 1928.

[93] Fridlyand Yu. S. On unremovable Carleman singularity for the Haar system. *Mat. zametki.*, 14:799–807, 1973. (in Russian)

[94] Gabisoniya O.D. On the best approximation of multivariate fucntions by polynomials with respect to the H.s. *Izv. Vyssh. Ucebn. Zaved. Matematica*, (1):30–46, 1968. (in Russian)

[95] Gaposhkin V.F. Unconditional bases of Orlicz spaces. *Uspehi Mat. Nauk*, 22:113–114, 1967. (in Russian)

[96] Gaposhkin V.F. On existence of unconditional basis of Orlicz spaces. *Funct. anal. i prilozen*, 1(4):26–32, 1967. (in Russian)

[97] Gaposhkin V.F. On unconditional bases of Orlicz spaces. *Sib. Mat. J.*, 9(2):280–287, 1968. (in Russian)

[98] Gaposhkin V.F. On the Haar system as an unconditional basis of $L_p(p > 1)$. *Mat. zametki*, 15(2):191–196, 1974. (in Russian)

[99] Garsia A. On a covex function inequality for martingales. *Ann. Probability*, 1:171–174, 1973.

[100] Garsia A. Topics in almost everywhere convergence. page 215, 1970.

[101] Gevorkyan G.G. On sets of uniqueness for subsequences of the H.s. *Uchenye zapiski Jerevan Univ.*, (3):15–19, 1983. (in Russian)

[102] Gevorkyan G.G. On u_p^*-sets for Haar, Walsh and trigonometric systems. *Izv. Acad. Nauk Arm. SSR*, 20(5):325–348, 1985. (in Russian)

[103] Gelbaum B.R. On the functions of Haar. *Ann. Math.*, 51:26–36, 1950.

[104] Golubov B.I. On Fourier-Haar series of continuous functions. *Dokl. Acad. Nauk SSSR*, 156(2):247–250, 1964. (in Russian)

[105] Golubov B.I. On Fourier-Haar series of continuous functions. *Izv. Acad. Nauk*, 28(6):1271–1296, 1964. (in Russian)

[106] Golubov B.I. On absolute convergence of Haar series. 20(5):198–202, 1965. (in Russian)

[107] Golubov B.I. On functions with bounded p-variation. *Uzv. Acad. Nauk. Ser. mat.*, 32(4):837–858, 1968. (in Russian)

[108] Golubov B.I. On Fourier-Haar series. *Itogi Nauki. Mat. anal.* Moscow, p.109–146, 1971. (in Russian)

[109] Golubov B.I., Efimov A.V., and Skvortsov V.A. *Walsh series and transforms. Theory and applications.* Moscow: Nauka, 1987. (in Russian)

[110] Gopinath B. and Kurahan R.P. Every T - space is equivalent to a T - space of continuous functions. *J. Approx. Theory*, 3:316–326, 1981.

[111] Gundy R. Decomposition of L_1 - bounded martingales. *Ann. Math. Statist*, 39:134–138, 1968.

[112] Guter R.S. and Ul'yanov P.L. On new results in theory of orthogonal series. *In the book: Kaczmarz S. and Steinhaus H. Theory of orthogonal series. Moscow, 1958.* p.335–456. (in Russian)

[113] Haar A. Zur Theorie der orthogonalen Functionensysteme. *Math. Annalen* 69:331–371, 1910.

[114] Hardy G.H., Littlewood J.E., and Polya G. *Inequalities.* Cambridge:Cambridge University Press, 1934.

[115] Holub J. R. On integral transforms of Haar functions. *Collog. Math.*, 59(2):23–39, 1990.

[116] Homutenko L.G. Exact estimates of Fourier-Haar coefficients of functions with bounded variation. *Mat. zametki*, 9(3):355–363, 1971. (in Russian)

[117] Homutenko L.G. Uniform approximation of functions of two variables by polynomials with respect to the Haar system. *Investigations on modern problems of summation and approximation of functions, and applications. Dnepropetrovsk*, 4:85–87, 1973. (in Russian)

[118] Horoshko N.P. On the best approximation in L-metric of some classes of functions by polynomials with respect to the Haar system. *Mat. zametki.*, 6(1):47–54, 1969. (in Russian)

[119] Horoshko N.P. and Lipovik V.V. Estimate of approximation of functions from $H_\omega[0,1]$ by Fourier-Haar sums in L_2-metric. *Investigations on moderm problems of summation and approximation of functions, and applications. Dnepropetrovsk*, p. 63–66, 1976. (in Russian)

[120] Horoshko N.P. Approximation of functions from $W^r H_\omega$ by Fourier-Haar sums in L_p-metric. *Approximations of functions by polynomials and splines, and summation of series. Dnepropetrovsk*, p. 65–68, 1990. (in Russian)

[121] Igari S. An extension of the interpolation theorem of Marcinkiewicz. *Tohoku Math. J.*, 15:343–358, 1963.

[122] Janson S. and Jones P.W. Integration between H^p spaces: The complex method. *J. Funct. Anal.*, 48:58–80, 1982.

[123] Jawerth B. and Milmon M. Extrapolation theory with application. 89(440):1–82, 1991.

[124] Johnson W.B. and Schechtman G. Martingale inequalities in rearrangement invariant fucntion spaces. *Isr. J. Math.*, 64(3):267–275, 1988.

[125] Kaczmarz S. and Steinhaus H. *Theory of orthogonal series.* Moscow:Fizmatgiz, 1958.

[126] Kadec V.M. and Popov M.M. On Schauder bases which are conditional in every hyperoctant. *Sib. Mat. J.*, 28(1):115–118, 1987. (in Russian)

[127] Kalton N.J. *Lattice structures on Banach Spaces.* Memoires AMS, N 493, 1993.

[128] Kantorovich L.V. and Akilov G.P. *Functional analysis.* Moscow:Nauka. 1977. (in Russian)

[129] Kashin B.S. and Saakyan A.A. *Orthogonal series.* Moscow:Nauka. 1984. (in Russian)

[130] Kaufman R. On the approximation of lacunary series by Brownian motion. *Acta Math. Acad. Sci. Hung.*, 35:61–66, 1980.

[131] Kazarian K.S. On the multiplicative complementation of basis sequences in $L_p, 1 \leq p < \infty$, up to bases of L_p. *Dokl. Acad. Nauk Arm. SSR*, 62(4):203–209, 1976. (in Russian)

[132] Kazarian K.S. On bases of $L_p, p \geq 1$, obtained by multiplication of minimal systems of function from L_p. *Dokl. Acad. Nauk Arm. SSR*, 64(3):151–155, 1977. (in Russian)

[133] Kazarian K.S. On the multiplicative complementation of some incomplete orthogonal systems up to bases of $L_p, 1 \leq p < \infty$. *Anal. Math.*, 4(1):37–52, 1978.

[134] Kazarian K.S. On bases and unconditional bases in the $L^p(d\mu), 1 \leq p < \infty$. *Stud. Math.*, 71(3):227–249, 1981.

[135] Kazarian K.S. On some questions of theory of orthogonal series. *Math. Sb.*, 119(2):278–294, 1982. (in Russian)

[136] Kemhadze G.G. On the convergence of spherical partial sums of multiple Fourier-Haar series. *Trudy Tbilisi Mat. inst. Acad. Nauk GSSR.*, 55:27–38, 1977. (in Russian)

[137] Kemhadze G.G. On the divergence of spherical partial sums of double Fourier-Haar series. *Nauchnye trudy Gruz. Polytechn. inst.*, (3/285):42–48, 1985. (in Russian)

[138] Ketonen T. On conditionality in L_s spaces. *Ann. Acad. Sci. Fenn. Ser. A.I. Diss 35*, p. 1–42, 1981.

[139] Kislovskaya N.M. and Osipov V.B. On the equivalence of rearrangements of the H.s. in L_1. *Mat. zametki*, 29(6):877–885, 1981. (in Russian)

[140] Komissarov A.A. On the equivalence of the Haar and the Franklin systems in r.i. spaces. *Uspehi Mat. Nauk.*, 37(2):203–204, 1982. (in Russian)

[141] Komissarov A.A. On the equivalence of the Haar and the Franklin systems in some function spaces. *Sib. Mat. J.*, 23(5):115–126, 1982. (in Russian)

[142] Komissarov A.A. On the equivalence of the Haar and the Franklin systems in r.i. spaces. *Trudy Saratovskoj Zimnej Shkoly, 1982*, Saratov. 2:65–68, 1983. (in Russian)

[143] Komissarov A.A. On some properties of the Haar and the Franklin systems. *Dokl. Acad. Nauk SSSR*, 285(6):1310–1313, 1985. (in Russian)

[144] Kopaliani T.S. On the convergence of multiple Fourier-Haar series in $H^p(\omega), H^p(\omega)$. *Soobscheniya Acad. Nauk GSSR.*, 135(2):245–247. (in Russian)

[145] Kracskowski A. An application of double moduli of variation to double Fourier-Haar series. *Funct. et approx.(PRL).*, 6:119–133, 1978.

[146] Kracskowski A. On double Haar series of functions of the class V_p. *Funct. et approxim.*, 5:25–30, 1977.

[147] Krantsberg A.S. On basis properties of the Haar system in the spaces with weights. *Trudy of the Moscow institute of electronic mechanical engineering.*, 24:14–21, 1972. (in Russian)

[148] Krantsberg A.S. On the rearrangements of the Haar system. *Mat. zametki*, 15:35–39, 1974. (in Russian)

[149] Krasnoselskii M.A. and Rutitskiy Ya. B. *Convex functions and Orlicz spaces.* Moscow: Fizmatgiz, 1958. (in Russian)

[150] Krein S.G., Petunin Yu.I., and Semenov E.M. *Interpolation of linear operators.* AMS. Providence, 1982.

[151] Krotov V.G. On the Haar series. *Sib. Mat. J.*, 14(1):111–127, 1973. (in Russian)

[152] Krotov V.G. Corrections to the paper "On the Haar series.". *Sib. Mat. J.*, 16(2):417–418, 1975. (in Russian)

[153] Krotov V.G. On the unconditional convergence of the Haar series in Λ^p_ω. *Mat. zametki*, 23(5):685–695, 1978. (in Russian)

[154] Kudajbergenov S.S. On the theorem of Jackson type for the H.s. in spaces with weights. *Trudy of Moscow institute of electronic technical engineering.*, 24:14–21, 1972. (in Russian)

[155] Kumtyshev P.S. On one generalization of the Haar functions. *Trudy of Ryazan radiotech. institute*, 67:146–149, 1975. (in Russian)

[156] Lindenstrauss J. and Pelczynski A. Contributions to the theory of the classical Banach spaces. *J. Functional Analysis*, 8:225–249, 1971.

[157] Linderstrauss J. and Tzafriri L. *Classical Banach Spaces I, Sequence Spaces.* Berlin: Springer, 1977.

[158] Linderstrauss J. and Tzafriri L. Classical Banach Spaces II, Function Spaces. Berlin: Springer, 1979.

[159] Liverani F. Una classe di nuclei in relazione con il sisteme ortonormale di Haar. *Atti. Semin. mat. fis. Univ. Modena.*, 14:157–168, 1965.

[160] Long Tran Duc and Triebel H. Equivalent norms end Schauder bases in anisotropic Besov spaces. *Proc. R. Soc. Edinburg*, (Sect. A 84):177–183, 1979.

[161] Lorentz G.G. Majorants in spaces of integrable functions. *Amer. J. Math.*, 77:484–492, 1955.

[162] Lunina M.A. On the set of points of unbounded divergence of the Haar series. *Vestnic MGU. Mat., mech.*, (4):13–20, 1976. (in Russian)

[163] Lunina M.A. On the set absolute convergence of the Haar series. *Anal. Math.*, 8(1):9–17, 1982.

[164] Luo Cheng. On Haar series. *Journ. Hang. Univ.*, (9):269–284, 1982.

[165] Luxemburg W.A.J. Rearrangement-invariant Banach function spaces. *Proc. Sympos. in Analysis. Queen's Papers in Pure and Appl. Math.*, 10:83–144, 1967.

[166] Marcinkiewicz J. *Collected Papers.* Warszawa: PWN., 1964.

[167] Marcinkiewicz J. Quelguess theorems sur less series orthogonales. *Ann.Soc.Polon.Math.*, 16:84–96, 1937.

[168] Maslov A.V. On the Fourier-Haar coefficients of the functions from L_p. *Mat. Sb.*, 126(4):490–514, 1985. (in Russian)

[169] Maslov A.V. On the Fourier-Haar series of functions of bounded variation. *Uspehi Mat. Nauk.*, 41(2):201–202, 1986. (in Russian)

[170] Matveev V.A. On the Fourier-Haar coefficients. *Izv. Vyss. Ucenbn. Zaved. Matematica*, (6):103–112, 1965. (in Russian)

[171] Matveev V.A. On the variation of functions and on the Fourier coefficients with respect to the Haar and Schauder systems. *Izv. Acad. Nauk. Ser. mat.*, 30(6):1397–1419, 1966. (in Russian)

[172] Maurey B. Isomorphismes entre espaces H_1. *Seminaire d'analyse fonctionnelle 1978-1979*, Number Exp. 19-20:1–7. Palaiseau, 1979.

[173] Maurey B. Systeme de Haar. *Seminaire Maurey - Schwartz*, Number Exp.I:1–11, 1974.

[174] McCullough S. and Wulbert D. The topological spaces that support Haar system. *Proc. Amer. Math, Soc.*, 14:678–692, 1985.

[175] McLaughlin J.R. Absolute convergence of Haar series. *Notices Amer. Math. Soc.*, 14(1):102, 1967.

[176] McLaughlin J.R. Haar series. *Trans. Amer. Math. Soc.*, 137:153–176, 1969.

[177] McLaughlin J.R. and Price J.J. Comparison of Haar series with gaps with trigonometric series. *Pacif. J. Math.*, 28(3):623–627, 1969.

[178] Meder J. On the absolute Norlund summability of orthogonal series for a system of H-type. *Colloq. Math.*, 18:203–211, 1967.

[179] Meyer Y. *Ondelettes et operateurs I.* Paris: Herman, 1990.

[180] Mikolas M. On the convergence of generalized Haar expansions. *Period. Politechn. Civ. Eng.*, 27(1-2):31–38, 1983.

[181] Mityagin B.S. Interpolation theorem for modular spaces. *Mat. Sb.*, 66(4):473–482, 1966. (in Russian)

[182] Mityagin B.S. Homotopic structure of linear group of Banach space. *Uspehi Mat. Nauk.*, 25(5):63–106, 1970. (in Russian)

[183] Morginthaler G. On Walsh - Fourier series. *Trans. A M S .*, 84:472–507, 1957.

[184] Morits F. On the unconditional convergence of the Haar series. *Izv. Acad. Nauk. SSSR. Ser. mat.*, 27(6):1229–1238, 1963. (in Russian)

[185] Movsisyan H.O. On the uniqueness of repeatedly convergent double Haar series. *Izv. Acad. Nauk Arm.SSR. Mat.*, 11(4):314–331, 1976. (in Russian)

[186] Movsisyan H.O. On the uniqueness of double Haar series converging with respect to the spheres. *Dokl. Acad. Nauk Arm.SSR*, 64(3):137–142, 1977. (in Russian)

[187] Muller R.F.X. A local version of a result of Gamlen and Gaudet. *Isr. J. Math.*, 63(2):212–222, 1988.

[188] Muller R.F.X. On subsequences of the Haar basis in $H_1(\delta)$ and isomorphism between H^1 - spaces. *Studia Math.*, 85(1):73–90, 1986 (87).

[189] Muller R.F.X. On the span of some three valued martingale difference sequences in L^p $(1 < p < \infty)$ and H^1. *Isr. J. Math.*, 60(1):39–53, 1987.

[190] Muller R.F.X. Permutation of the Haar system. *Lect. Notes. Math.*, 1469:125–126. 1991.

[191] Mushegyan G.M. On the sets of uniqueness for the Haar system. *Izv. Acad. Nauk Arm.SSR. Mat.*, 2(6):360–361, 1967. (in Russian)

[192] Mushegyan G.M. On the sets of uniqueness for the Haar system. *Izv. Acad. Nauk Arm.SSR. Mat.*, 4(1):55–65, 1969. (in Russian)

[193] Mushegyan G.M. On the coefficients of everywhere convergent Haar series with rearranged terms. *Izv. Acad. Nauk Arm.SSR. Math*, 11(4):276–300, 1978. (in Russian)

[194] Musielak J. On generalized Orlicz sequence space of Fourier coefficients for Haar and trigonometric system. *Colloq.Math.Soc.Janos Boyai (Alfred Haar Mem. Conf., Budapest, Hung., 1985)*, 49:64–649, 1987.

[195] Musilimov B. and Gusipov K. On the unconditional convergence of the Fourier series with respect to the system of Haar type., *V Soviet-Czechoslovakian meeting on the applications of the methods of the theory of functions and functional analysis to the problems of mathematical physics. Alma-Ata., 1976*, Novosibirsk:93-96, 1979. (in Russian)

[196] Neveu J. *Bases mathematiques du calcul des brobabilites.* Paris: Masson et Cie, 1964.

[197] Neveu J. *Discrete parameter martingales.* Amsterdam, 1975.

[198] Nikishin E.M. and Ul'yanov P.L. On the absolute unconditional convergence. *Uspehi Mat. Nauk.*, 22(3):240–242, 1967. (in Russian)

[199] Novikov I. Ya. On the subsequences of the Haar system in L_1. *Uspehi Mat. Nauk*, 39(1):121–122, 1984. (in Russian)

[200] Novikov I. Ya. Subspaces of L_1 generated by subsequences of the Haar system. *Operator equations in function spaces.*, Voronezh:60–71, 1986. (in Russian)

[201] Novikov I. Ya. Inequality of B.Davis in rearrangement invariant spaces, *XIII school on the theory of operators in function spaces.* Kujbyshev:139-140, 1988. (in Russian)

[202] Novikov I. Ya. Martingale inequalities in Rearrangement Invariant Function Spaces. *Teubner-texte Mathematik.* 120:120–127, 1991.

[203] Novikov I. Ya. Criterian of equivalence of the Haar and the Franklin in R.I. spaces. *Mat. zametki.*, 52(3):96–101, 1992. (in Russian)

[204] Novikov I.Ya. and Semenov E.M. On the Fourier-Haar coefficients. *Mat. zametki.*, 36(3):351–358, 1984. (in Russian)

[205] Ogava S. Une remarque sur l'approximation de l'integrale stochestigue du type noncausal par une Suite des integrales de Stiltjes. *Tohoku Math. J. Ser. II.*, 36:41–48, 1984.

[206] Ohkita M. and Kobayashi Y. Solution of time-varying linear delay system by rationalized Haar functions. *IMACS Ann. Comput. and Appl. Math., Pt. 2*, 1:693–698, 1989.

[207] Olevskii A.M. On one orthonormal system and its application. *Mat. Sb.*, 71(3):207–336, 1966. (in Russian)

[208] Olevskii A.M. Fourier series and Lebesque fucntion. *Uspehi Mat. Nauk.*, 22:236–239, 1967. (in Russian)

[209] Olevskii A.M. *Fourier series with respect to general orthogonal system.* Berlin: Springer, 1975.

[210] Orlicz W. Uder eine gewisse Klasse von Raumen von Types B. *Bull. Acad. Polon. Sci.*, 3:207 220, 1932.

[211] Osipov V.B., Sadovskaya O.B., and Semenov E.M. Two-sided estimates of Fourier-Haar multipliers. *Dokl. Acad. Nauk USSR.*, 294(2):276–278, 1987. (in Russian)

[212] Osvald P. l^p approximation durch Reihen nach dem Haar-Orthogonal-System und dem Faber-Schauder-System. *J. Approxim. Theory.*, 33(33):1–27, 1981.

[213] Oswald P. On some convergence properties of Haar-Fourier series in the class $\varphi(l)$. *Acta. Math. Hung.*, 42(4):279–293, 1983.

[214] Otyrba D. On rearrangements of Haar system in BMO. *Uspehi Mat. Nauk* 49(3):183–184, 1994 (in Russian)

[215] Ovsepyan R.I. On some theorems of uniqueness for the Haar system and centralized ones of continuous fucntions. *Izv. Acad. Nauk Arm.SSR. Ser. mat.*, 25,5:421–431, 1990. (in Russian)

[216] Pal L.G. and Schipp F. On Haar schauder series. *Acta. Sci. Math.*, 3(1-2):53–58, 1970.

[217] Pal L.G. and Schipp F. On the steinhaus conjecture with respect to the Haar series. *Proc. Conf. Constructive Theory of Fucntions. Budapest, 1969.* Budapest: Academiai Kiado, 1972, p.343–349.

[218] Paley R. A remarkable series of orthogonal functions I. *Proc. Lond. Math. Soc.*, 34:241–264, 1932.

[219] Pelczynski A. Norm of classical operators in functional spaces. *Asterisque.*, 31:137–162, 1985.

[220] Pelszynski A. Projections in certain Banach spaces. *Studia Math.*, 19:209–228, 1960.

[221] Pelszanski A. and Singer I. On non-equivalent basis and conditional bases in Banach Spaces. *Studia Math.*, 25, 1964.

[222] Petrovskaya M.B. Some theorems of uniqueness for series with respect to the Haar system. *Vestnik MGU. Ser. mat., mech.*, (56):15–28, 1964. (in Russian)

[223] Petrovskaya M.B. On zero-series with respect to the Haar system and the sets of uniquence. *Izv. Acad. Nauk USSR. Ser. mat.*, 28(4):773–798, 1964. (in Russian)

[224] Petrovskaya M.B. On the series with respect to the Haar system and functions of class H^1_ω. *Sib. Math. J.*, 9(4):863–879, 1968. (in Russian)

[225] Plichko A.M. and Tokarev E.V. On the bases of rearrangement invariant spaces. *Mat. zametki.*, 29(6):227–234, 1987. (in Russian)

[226] Plichko A.M. and Popov M.N. Symmetric function spaces on atomless probability spaces. *Dissertationes Mathematical (Rozprawey Matematyczne).*, 306:1–85, 1990.

[227] Pogosyan N.B. Summation of the Haar and Walsh series. *Uspehi Mat. Nauk.*, 35(2):219–220, 1980. (in Russian)

[228] Popa N. Interpolation series for rearrangement invariant p spaces of functions, $0 < p < 1$, and some application. *Rend. Circ. Mat. Palermo, Ser II.*, 2:199–216, 1982.

[229] Pribegin S. G. On absolute convergence of the Foureir-Haar series. *Izv. Vyss. Ucebn. Zaved. Matematica.*, (8):77–80, 1981. (in Russian)

[230] Price J.J. Haar series and adjustments of fucntions on small sets. *Jllinois J. Math.*, 14:82–87, 1970.

[231] Price J.J. and Zink R.E. On sets of completeness for famillies of Haar functions. *Trans. Amer. Math. Soc.*, 119(2):262–269, 1965.

[232] Prohorenko V.I. On divergent series with respect to the Haar system. *Izv. Vyssh. Uceb. Zaved. Matematica.*, (1):62–68, 1971. (in Russian)

[233] Renault J. Representations des products croises d'algebras de groupoides. *J. Oper. Theory.*, 18:67–97, 1987.

[234] Robakidze M.G. Fourier-Haar coefficients of absolutely continuous functions. *Soobscheniya Acad. Nauk GSSR.*, 133(2):245–247, 1989. (in Russian)

[235] Rodin V.A. and Semyonov E.M. Rademacher series in symmetric spaces. *Anal. Math.*, 1:207–222, 1975.

[236] Ryff J.V. Orbits of L^1-fucntions under doubly stochastic transformations. *Trans. Amer. Math. Soc.*, 117:92–100, 1965.

[237] Sadovskaya O.B. Criterion of boundedness of one operator of rearrangement of the Haar system. *Technical Report 4342-83, Dep. VINITI 05.08.1983.* (in Russian)

[238] Sadovskaya O.B. On the question of boundedness of Fourier-Haar multiplier. *Technical Report 8882, Dep. VINITI 25.12.1986.* (in Russian)

[239] Sadovskaya O.B. and Semenov E.M. Fourier-Haar multipliers, *Dokl. Sem. Vekua Inst. Prikl. Mat., Tbilisi,* 1988, p.126–128. (in Russian)

[240] Schauder J. Einige Eigenschaft des Haarschen Orthogonalsystems. *Math. Zeit.*, 28:315–329, 1928.

[241] Schechtmen G. Remark on unconditional basic sequences in L_p $(1 < p < \infty$. *Isr. J. Math.*, 19:220–224, 1974.

[242] Semenov E.M. On the equivalence in L_p of rearrangements of the Haar system. *Dokl. Acad. Nauk SSSR.*, 242(6):1258–1260, 1978. (in Russian)

[243] Semenov E.M. On the boundedness of operators rearranging the Haar system in the space L_p. *Integral Equation and Operator Theory.*, 6(3):385–404, 1983.

[244] Semenov E.M. Maximal Calderon operator. *Dokl. Acad. Nauk SSSR.*, 315(1):31–34, 1990. (in Russian)

[245] Semenov E.M. Estimates for operator of weak type. pages *Proc. Int. Symp. Sapporo. Japan. 31 Aug - 4 Sep., 1990. Function analysis and related topics.* Singapore, 1991, p. 172–178.

[246] Semenov E.M. Basic properties of the Olevskii system. *Mat.zametki* 54(3):155–156, 1993 (in Russian)

[247] Semenov E.M. Haar system's rearrangements in the Lorentz spaces. *Sib.Mat J.* 34(6):158–164, 1993. (in Russian)

[248] Semenov E.M. Estimates of Haar multipliers in the space L_1. *Dokl. Rus. Akad. Nauk*, 334(5):564–565, 1994 (in Russian)

[249] Semenov E.M. and Shteinberg A.M. Estimates of norms of operator blocks in Banach lattices. *Mat. Sb.*, 126(3):327–343, 1985. (in Russian)

[250] Semenov E.M. and Stockert B. Haar system rearrangement in the space L_p. *Analysis Math.*, 7(4):277–298, 1981.

[251] Semenov E.M. and Tsirelson B.S. Problem of smallness of operator blocks in L_p spaces. *Zeit. fur Anal. and inre Anwend.*, 2(4):367–573, 1983.

[252] Shaginyan L.A. On the summation to $+\infty$ of the Haar series by the methood $(C, 1)$. *Izv. Acad. Nauk Arm.SSR.*, 16:366–384, 1981. (in Russian)

[253] Sharaputdinov I.I. On the basis properties of the Haar system in spaces $L^{p(t)}([0,1])$ and the localization in mean principle. *Mat. Sb.*, 130(2):275–283, 1986. (in Russian)

[254] Shergin Yu. V. On the Fourier projections on the linear spans of subsequences of the Haar system in the space $L^1[0,1]$. *Vestnik LGU.*, (1):118–120, 1989. (in Russian)

[255] Shekhtman B. On projections in approximation theory. *Approx. Theory and spline funct. Proc. NATO Adv. Study. Inst.*, St. John's., 1984, p. 455–466.

[256] Shimogaki T. A note on norms of compression operators of function spaces. *Proc. Japan. Acad.*, 46(3):239–242, 1970.

[257] Shimogaki T. Hardy - Littlewood majorants in functional spaces. *J. Math. Soc. Japan.*, 17:365–373, 1965.

[258] Shipp F. On a generalization of the Haar system. *Acta Math. Acad. Sci. Hunger.*, 33:83–188, 1979.

[259] Shipp F. On equivalence of rearrangements of the Haar system in dyadic Hardy and BMO spaces. *Analysis Math.*, 16(2):35–142, 1990.

[260] Shipp F. and Simon P. Invastigation of Haar and Franklin series on Hardy spaces. *Anal. Math.*, 8(1):47–56, 1982.

[261] Shipp F., Wade W.R., Simon P., and Pal J. *Walsh series: An introduction to diadic Harmonic analysis.* Budapest: Acad. Kiado., 1990.

[262] Shirey J. The convergence of series for various choices of sign in Banach spaces. *Can. J. Math.*, 27(2):475–480, 1975.

[263] Sjolin P. Tha Haar and Franklin systems are not equivalenet in L_1. *Bull. Acad. Pol. Sci. Ser. sci. math., astron. et phys.*, 25(11):1099–1100, 1977.

[264] Skvortsov V.A. Cantor type theorem for the Haar system. *Vestnik MGU. Ser. math., mech.*, (5):12–28, 1964. (in Russian)

[265] Skvortsov V.A. Calculation of coefficients of everywhere convergent Haar series. *Mat. Sb.*, 75(3):349–360, 1968. (in Russian)

[266] Skvortsov V.A. Differentiation with respect to the nets and the Haar series. *Mat. zametki.*, 4(1):33–40, 1968. (in Russian)

[267] Skvortsov V.A. On the Haar series converging with respect to the subsequences of partial sums. *Dokl. Acad. Nauk SSSR.*, 183(4):784–786, 1968. (in Russian)

[268] Skvortsov V.A. On the uniqueness of the Haar series converging with respect to subsequences of partial sums. *Mat. zametki.*, 4(6):707–714, 1968. (in Russian)

[269] Skvortsov V.A. Uniqueness theorems for the summation methods of the Haar series. *Mat. zametki.*, 9(4):449–458, 1970. (in Russian)

[270] Skvortsov V.A. On one example of double Haar series. *Mat. zametki.*, 28(3):343–353, 1980. (in Russian)

[271] Skvortsov V.A. On uniqueness theorem for double Haar series with convergent spherical partial sums. *Vestnik MGU. Mat., mech.*, (1):12–17, 1981. (in Russian)

[272] Skvortsov V.A. Some properties of double Haar series with everywhere convergent spherical partial sums. *Mat. zametki.*, 33(1):89–95, 1983. (in Russian)

[273] Skvortsov V.A. In connection with divergent Haar series. *Mat. zametki.*, 36(4):509–516, 1984. (in Russian)

[274] Skvortsov V.A. Questions of uniqueness of representation of functions by multiple Haar and Walsh series. *Theory of approximation of functions: Trudy International conference on approximation theory. Kiev, 31 May - 5 June.*, 1987, p. 402–403. (in Russian)

[275] Skvortsov V.A. On one theorem of uniqueness for multiple Haar series. *Izv. Acad. Nauk Arm SSR.*, 23(3):293–296, 1988. (in Russian)

[276] Skvortsov V.A. and Talalyan A.A. Some questions of uniqueness for multiple series with respect to the Haar and the trigonometric systems. *mat. zametki.*, 46(2):104–113, 1989. (in Russian)

[277] Smailov E.S. Imbedding theorems of different metric for function spaces with the orthonormal Haar basis, *Math. Issled.*, 2:17–21, Karaganda, 1975 (in Russian)

[278] Soares de Sowra G. The dyadic special atom space. *Proc. Conf. Minneapolis. Lect. Notes. Math.*, 908:297–299, 1982.

[279] Sobol I.M. Multivariate functions with fast convergent Haar series. *Dokl. Acad. Nauk SSSR.*, 132(4):773–776, 1960. (in Russian)

[280] Sobol I.M. On the application of the Haar series in quadrature formuleas, *Questions of numerical math. and computers*, Moscow, 1963, p. 31–35. (in Russian)

[281] Sobol I.M. On the method of the Haar series in the theory of multivariable quadratures. *Abstracts of short scientific talks at the International Math. congress. Section 14*, Moscow, 1966, p.48. (in Russian)

[282] Sobol I.M. Application of the haar series for error estimates of calculation of infinite dimensional integrals. *Dokl. Acad. Nauk SSSR.*, 175(1):34–37, 1967. (in Russian)

[283] Sobol I.M. *Multivariate quadrature formulas and the Haar functions.* Moscow: Nauka, 1969. (in Russian)

[284] Solomyak M.Z. On the orthogonal basis in the Banach space. *Vestnik LGU.*, (1):27–36, 1957. (in Russian)

[285] Splettstoesser W. and Ziegler W. A generalized Haar functions, the Haar transform of R^+, and the Haar sampling theorem. *Colloq.Math.Soc.Janos Boyai (Alfred Haar Mem. Conf., Budapest, Hung., 1985)*, 49:873–896, 1987.

[286] Splettstosser W. and Wagner H.J. Eihe Dyadische infinitesimalrechnung fur Haar-Funktionen. *Z. Angew. Math. Mech.*, 57:527–540, 1977.

[287] Storogenko E.A., Krotov B.G., and Oswald P. Direct and inversc Jackson type theorems in L_p spaces, $0 < p < 1$. *Mat. Sb.*, 98(3):395–415, 1975. (in Russian)

[288] Szelmeczka J. On absolute convergence of multiple Haar and Rademacher series. *Math. Prace. Mat.*, 17:475–479, 1974.

[289] Szokefalvi Nagy B. Approximation properties of orthogonal expansions. *Acta. Sci. math.*, 15:31–37, 1953/54.

[290] Takayama Y. Estimates on the maximum of partial sums of Haar series. in relation to Ulijanov's theorem. *Mem. Miyakonojo Techn. Coll.*, (11):27–32, 1976 (1977).

[291] Takayama Y. On convergence of Haar series. *Mem. Miyakonojo Techn. Coll. Res. Rept.*, (12):1–4, 1977 (1978).

[292] Talayan A.A. On subsequences of the Haar system. *Dokl. Acad. Nauk Arm. SSR.*, 55(1):3–6, 1972. (in Russian)

[293] Talayan A.A. On the rate of convergence to zero of the zero-series with respect to the Haar system. *Dokl. Acad. Nauk Arm. SSR.*, 77:16–19, 1983. (in Russian)

[294] Talayan A.A. On the uniqueness of the Haar series which converge in the metric of $L_p[0,1], 0 < p < 1$, and in measure. *Mat. Sb.*, 126(1):101–114, 1985. (in Russian)

[295] Talayan A.A. and Arutyunyan F.G. On the convergence of series with respect to the Haar system. *Mat. Sb.*, 66(2):240–247, 1965. (in Russian)

[296] Tazabekov S. On the transformed Fourier series with respect to the system of the Haar type. *Technical Report 8705-B88, Dep. VINITI 01.12.1988.* (in Russian)

[297] Tazabekov S. On the Fourier coefficients of the functions from L_p with respect to the system of the Haar type, *Modern questions of the theory of functions and of the functional analysis*, Karaganda, 1988, p. 109–118. (in Russian)

[298] Tazabekov S. On the imbedding of the classes $E_p(\lambda)$ with respect to the system of the Haar type in the Marcinkiewicz space. *Vestnik Acad. Nauk Kaz.SSR.*, (12):53–56, 1989. (in Russian)

[299] Timan M.F. and Gajmnazarov G. The best approximations and the absolute convergence of the Fourier-Haar series. *Dokl. Acad. Nauk SSSR.*, 198(6):1280–1282, 1971. (in Russian)

[300] Timan M.F. and Tuhliev K. Properties of some orthonormal systems. *Tzv. Vyss. Uceb. Zaved. Matematica.*, (9), 1985. (in Russian)

[301] Tkebuchava G.E. On the multiple Fourier-Haar series. *Some questions of the theory of functions*,1:307–331. Tbilisi, 1979. (in Russian)

[302] Tkebuchava G.E. On unconditional basis properties of the Haar system in the spaces with weights. *Soobscheniya Acad. Nauk GSSR.*, 102(2):285–287, 1981. (in Russian)

[303] Tkebuchava G.E. On the multiple systems of functions in the Orlicz spaces. *Soobcheniya Acad. Nauk GSSR.*, 129(2):249–252, 1988.

[304] Tkebuchava G.E. On the unconditional bases of function spaces. *Some questions of the theory of functions and of the functional analysis.*, 4:207–224, Tbilisi, 1988. (in Russian)

[305] Riebel H. On Haar bases in Besov spaces. *Serdica*, 4(4):330–343, 1978.

[306] Triebel H. *Interpolation theory, function spaces, differential operators.* Berlin, 1978.

[307] Triebel H. Spline bases and spline representations in function spaces. *Arch. Math.*, 36:348–359, 1981.

[308] Tsagerejshvili V.Sh. Absolute convergence of orthogonal Fourier series. *Some questions of the theory of functions.*, Tbilisi, 1981, p. 93–109. (in Russian)

[309] Tutai E. Some observations concerning the class unconditionallike basic sequence. *Bull. Pol. Acad. Sci. Math.*, 35:35–42, 1987.

[310] Ul'yanov P.L. On some solved and unsolved series. volume 2, pages 694–704. Trudy of the 4-th All-Union mathematical Congress., 1961. in Russian.

[311] Ul'yanov P.L. Divergent Fourier series. *Uspehi Mat. nauk.*, 16(3):61–142, 1961. in Russian.

[312] Ul'yanov P.L. Divergent series with respect to the Haar system and with respect to the bases. *Dokl. Acad. Nauk SSSR.*, 138(3):556–559, 1961. in Russian.

[313] Ul'yanov P.L. On the Weyl multipliers fot the unconditional convergence. *Mat. Sb.*, 60(1):39–62, 1963. in Russian.

[314] Ul'yanov P.L. On the Haar series. *Dokl. Acad. Nauk SSSR.*, 149(3):532–534, 1963. in Russian.

[315] Ul'yanov P.L. Solved and unsolved problems of the theory of trigonometric and orthogonal series. *Uspehi Mat. nauk.*, 19(1):3–69, 1964. in Russian.

[316] Ul'yanov P.L. On the Haar series with monotone coefficients. *Izv. Acad. Nauk SSSR, Ser. mat.*, 28(4):925–950, 1964. in Russian.

[317] Ul'yanov P.L. On the Haar series. *mat. Sb.*, 63(3):356–391, 1964. in Russian.

[318] Ul'yanov P.L. The Haar series. *Vestnik MGU. Mat., mech.*, (4):35–43, 1965. in Russian.

[319] Ul'yanov P.L. *The Haar series. Invastigations of the modern problems of the constructive theory of functions.*, pages 397–406. Baku, 1965. in Russian.

[320] Ul'yanov P.L. On transformations of Fourier-Haar coefficients. 48(3):478–484, 1994.

[321] Ul'yanov P.L. On some properties of the Haar series. *Mat. zametki.*, 1:17–24, 1967. in Russian.

[322] Ul'yanov P.L. Absolute convergence of Fourier-Haar series for superpositions of functions. *Anal. Math.*, 4:225–236, 1978.

[323] Ul'yanov P.L. On the uniqueness of the Haar series with monotone coefficients. *Vestnik MGU. Mat., mech.*, 38(6):74–85, 1983. in Russian.

[324] Ul'yanpov P.L. Superpositions of functions and Fourier coefficients. *Trudy MIAN.*, 172:338–348, 1985. in Russian.

[325] Uljanov R.L. Haar series and related questions. volume 49, pages 57–95. Alfred Haar Mem. Conf., Budapest / Hung., 1985.

[326] Valashek Ya. Imbedding theorems and inequalities of different metrics for the best approximation with respect to the Haar system. *Izv. Vyss. Uceb. Zaved. Matematica.*, (4):11–15, 1980. in Russian.

[327] Vlasova E.A. Convegence almost everywhere with respect to the system of the Haar type. pages 72–73. Materialy 13 conferencii uchenyh mech.-mat. fakul'teta MGU. M., 1987. in Russian.

[328] Vlasova E.A. On one class of orthogonal system of convergence. *Mat. zametki.*, 43(6):734–745, 1988. in Russian.

[329] Vlasova E.A. On the Haar series. *Izv. Vyss. Uceb. Zaved. Matematica.*, (9):3–14, 1990. in Russian.

[330] Wade W.R. A unifide approch to uniqueness of Walsh series and Haar series. *Proc. Amer. Math. Soc.*, 39(1):61–65, 1987.

[331] Wade W.R. A uniqueness theorem for Haar and Walsh series . *Trans. Amer. Math. Soc.*, 141:187–194, 1969.

[332] Wade W.R. Haar and Walsh Fourier series of Perron integrable functions. *J. Indian Math. Soc.*, 38(1-4):19–35, 1974(1975).

[333] Wade W.R. *Uniquenese theory of the Haar and Walsh systems.* PhD thesis, Doct. diss Riverside. Univ. Calif., 1968.

[334] Wade W.R. Recent developments in the theory of Haar series. *Colloquim Mathematicum.*, 52(2):213–238, 1987.

[335] Wang Si. lei. On the Haar series. *Chiness Math.*, 8(2):222–233, 1966.

[336] Watari C. A generalization of Haar function. *Tohoku Math. J.*, 8(3):286–289, 1956.

[337] Watari C. Mean convergence of Walsh-Fourier series. *Tohoku Math. J.*, 16(2):183–188, 1964.

[338] Watari C. Multipliers for Walsh-Fourier series. *Tohoku Math. J.*, 16(3):239–251, 1964.

[339] Welker G. Haarsche Systeme von Komplexen Exponentialsummen. *Arch. Math.*, 33:375–378, 1980.

[340] Wierzbicka I. The absolute summability of Haar series by some matrix method. *Bull. Acad. pol. sci. Ser. sci. math., astron. et phys.*, 21(5):405–412, 1973.

[341] Yano S. On a lemma of Martinkiewicz and its applications to Fourier series. *Tohoku Math. J.*, 11:191–215, 1959.

[342] Zaderej P.V. and Zaderei N.N. *Uniform approximation of the multivariate continuous functions by polynomials with respect to the Haar system. Theory of approximation of functions and applications.*, pages 60–66. Kiev, 1984. in Russian.

[343] Zerekidze T.Sh. Convergence of multiple Fourier-Haar series and strong differentiability of integrals. *Trudy Tbilisi mat. inst. Acad. Nauk GSSR.*, 76:80–99, 1985. in Russian.

[344] Zink R. An orthonormal basis for $C[0,1]$ that is not an unconditional basis for $L^p[0,1], 1 < p \neq 2$. *Proc. AMS.*, 97(1):33–37, 1986.

[345] Zotikov S.V. On the convergence almost everywhere of the series with respect to the system of the Haar type. *Mat. J.*, 14(4):760–765, 1973. in Russian.

[346] Zotikov S.V. On the absolute convergence of the series with respect to the system of the Haar type. *Izv. Vyss. Ucebn. Zaved. Matematica.*, (11):31–43, 1974. in Russian.

[347] Zotikov S.V. *On the convergence of the Fourier-Haar integrals. Applications of the functional analyses to the approximation theory.*, pages 81–89. Kalinin, 1982. in Russian.

[348] Zotikov S.V. *On the rate of decreasing of the Fourier-Haar transform. Applications of the functional analyses to the approximation theory.*, pages 46–54. Kalinin, 1984. in Russian.

[349] Zotikov S.V. *On the representation of functions by the Fourier-Haar integrals. Applications of the functional analyses to the approximation theory.*, pages 52–57. Kalinin, 1986. in Russian.

[350] Zotikov S.V. *Fourier-Haar transform and integral in the space L_2. Applications of the functional analyses to the approximation theory.*, pages 42–49. Kalinin, 1987. in Russian.

[351] Zygmund A. On a theorem of Martinkiewicz Concerning Interpolation of Operators. *J. Math. Pures Appl.*, 35(2):223–248, 1956.

[352] Zygmund A. *Trigonometric series.*, volume 2. Cambridge University Press., 1959.

Other *Mathematics and Its Applications* titles of interest:

A.F. Filippov: *Differential Equations with Discontinuous Righthand Sides*. 1988, 320 pp. ISBN 90-277-2699-X

A.T. Fomenko: *Integrability and Nonintegrability in Geometry and Mechanics*. 1988, 360 pp. ISBN 90-277-2818-6

G. Adomian: *Nonlinear Stochastic Systems Theory and Applications to Physics*. 1988, 244 pp. ISBN 90-277-2525-X

A. Tesar and Ludovt Fillo: *Transfer Matrix Method*. 1988, 260 pp. ISBN 90-277-2590-X

A. Kaneko: *Introduction to the Theory of Hyperfunctions*. 1989, 472 pp. ISBN 90-277-2837-2

D.S. Mitrinovic, J.E. Pecaric and V. Volenec: *Recent Advances in Geometric Inequalities*. 1989, 734 pp. ISBN 90-277-2565-9

A.W. Leung: *Systems of Nonlinear PDEs: Applications to Biology and Engineering*. 1989, 424 pp. ISBN 0-7923-0138-2

N.E. Hurt: *Phase Retrieval and Zero Crossings: Mathematical Methods in Image Reconstruction*. 1989, 320 pp. ISBN 0-7923-0210-9

V.I. Fabrikant: *Applications of Potential Theory in Mechanics. A Selection of New Results*. 1989, 484 pp. ISBN 0-7923-0173-0

R. Feistel and W. Ebeling: *Evolution of Complex Systems. Selforganization, Entropy and Development*. 1989, 248 pp. ISBN 90-277-2666-3

S.M. Ermakov, V.V. Nekrutkin and A.S. Sipin: *Random Processes for Classical Equations of Mathematical Physics*. 1989, 304 pp. ISBN 0-7923-0036-X

B.A. Plamenevskii: *Algebras of Pseudodifferential Operators*. 1989, 304 pp. ISBN 0-7923-0231-1

N. Bakhvalov and G. Panasenko: *Homogenisation: Averaging Processes in Periodic Media. Mathematical Problems in the Mechanics of Composite Materials*. 1989, 404 pp. ISBN 0-7923-0049-1

A.Ya. Helemskii: *The Homology of Banach and Topological Algebras*. 1989, 356 pp. ISBN 0-7923-0217-6

M. Toda: *Nonlinear Waves and Solitons*. 1989, 386 pp. ISBN 0-7923-0442-X

M.I. Rabinovich and D.I. Trubetskov: *Oscillations and Waves in Linear and Nonlinear Systems*. 1989, 600 pp. ISBN 0-7923-0445-4

A. Crumeyrolle: *Orthogonal and Symplectic Clifford Algebras. Spinor Structures*. 1990, 364 pp. ISBN 0-7923-0541-8

V. Goldshtein and Yu. Reshetnyak: *Quasiconformal Mappings and Sobolev Spaces*. 1990, 392 pp. ISBN 0-7923-0543-4

Other *Mathematics and Its Applications* titles of interest:

I.H. Dimovski: *Convolutional Calculus.* 1990, 208 pp. ISBN 0-7923-0623-6

Y.M. Svirezhev and V.P. Pasekov: *Fundamentals of Mathematical Evolutionary Genetics.* 1990, 384 pp. ISBN 90-277-2772-4

S. Levendorskii: *Asymptotic Distribution of Eigenvalues of Differential Operators.* 1991, 297 pp. ISBN 0-7923-0539-6

V.G. Makhankov: *Soliton Phenomenology.* 1990, 461 pp. ISBN 90-277-2830-5

I. Cioranescu: *Geometry of Banach Spaces, Duality Mappings and Nonlinear Problems.* 1990, 274 pp. ISBN 0-7923-0910-3

B.I. Sendov: *Hausdorff Approximation.* 1990, 384 pp. ISBN 0-7923-0901-4

A.B. Venkov: *Spectral Theory of Automorphic Functions and Its Applications.* 1991, 280 pp. ISBN 0-7923-0487-X

V.I. Arnold: *Singularities of Caustics and Wave Fronts.* 1990, 274 pp. ISBN 0-7923-1038-1

A.A. Pankov: *Bounded and Almost Periodic Solutions of Nonlinear Operator Differential Equations.* 1990, 232 pp. ISBN 0-7923-0585-X

A.S. Davydov: *Solitons in Molecular Systems. Second Edition.* 1991, 428 pp. ISBN 0-7923-1029-2

B.M. Levitan and I.S. Sargsjan: *Sturm-Liouville and Dirac Operators.* 1991, 362 pp. ISBN 0-7923-0992-8

V.I. Gorbachuk and M.L. Gorbachuk: *Boundary Value Problems for Operator Differential Equations.* 1991, 376 pp. ISBN 0-7923-0381-4

Y.S. Samoilenko: *Spectral Theory of Families of Self-Adjoint Operators.* 1991, 309 pp. ISBN 0-7923-0703-8

B.I. Golubov A.V. Efimov and V.A. Scvortsov: *Walsh Series and Transforms.* 1991, 382 pp. ISBN 0-7923-1100-0

V. Laksmikantham, V.M. Matrosov and S. Sivasundaram: *Vector Lyapunov Functions and Stability Analysis of Nonlinear Systems.* 1991, 250 pp. ISBN 0-7923-1152-3

F.A. Berezin and M.A. Shubin: *The Schrödinger Equation.* 1991, 556 pp. ISBN 0-7923-1218-X

D.S. Mitrinovic, J.E. Pecaric and A.M. Fink: *Inequalities Involving Functions and their Integrals and Derivatives.* 1991, 588 pp. ISBN 0-7923-1330-5

Julii A. Dubinskii: *Analytic Pseudo-Differential Operators and their Applications.* 1991, 252 pp. ISBN 0-7923-1296-1

V.I. Fabrikant: *Mixed Boundary Value Problems in Potential Theory and their Applications.* 1991, 452 pp. ISBN 0-7923-1157-4

Other *Mathematics and Its Applications* titles of interest:

A.M. Samoilenko: *Elements of the Mathematical Theory of Multi-Frequency Oscillations.* 1991, 314 pp. ISBN 0-7923-1438-7

Yu.L. Dalecky and S.V. Fomin: *Measures and Differential Equations in Infinite-Dimensional Space.* 1991, 338 pp. ISBN 0-7923-1517-0

W. Mlak: *Hilbert Space and Operator Theory.* 1991, 296 pp. ISBN 0-7923-1042-X

N.Ja. Vilenkin and A.U. Klimyk: *Representation of Lie Groups and Special Functions. Volume 1: Simplest Lie Groups, Special Functions, and Integral Transforms.* 1991, 608 pp. ISBN 0-7923-1466-2

N.Ja. Vilenkin and A.U. Klimyk: *Representation of Lie Groups and Special Functions. Volume 2: Class I Representations, Special Functions, and Integral Transforms.* 1992, 630 pp. ISBN 0-7923-1492-1

N.Ja. Vilenkin and A.U. Klimyk: *Representation of Lie Groups and Special Functions. Volume 3: Classical and Quantum Groups and Special Functions.* 1992, 650 pp. ISBN 0-7923-1493-X
(Set ISBN for Vols. 1, 2 and 3: 0-7923-1494-8)

K. Gopalsamy: *Stability and Oscillations in Delay Differential Equations of Population Dynamics.* 1992, 502 pp. ISBN 0-7923-1594-4

N.M. Korobov: *Exponential Sums and their Applications.* 1992, 210 pp.
ISBN 0-7923-1647-9

Chuang-Gan Hu and Chung-Chun Yang: *Vector-Valued Functions and their Applications.* 1991, 172 pp. ISBN 0-7923-1605-3

Z. Szmydt and B. Ziemian: *The Mellin Transformation and Fuchsian Type Partial Differential Equations.* 1992, 224 pp. ISBN 0-7923-1683-5

L.I. Ronkin: *Functions of Completely Regular Growth.* 1992, 394 pp.
ISBN 0-7923-1677-0

R. Delanghe, F. Sommen and V. Soucek: *Clifford Algebra and Spinor-valued Functions. A Function Theory of the Dirac Operator.* 1992, 486 pp.
ISBN 0-7923-0229-X

A. Tempelman: *Ergodic Theorems for Group Actions.* 1992, 400 pp.
ISBN 0-7923-1717-3

D. Bainov and P. Simenov: *Integral Inequalities and Applications.* 1992, 426 pp.
ISBN 0-7923-1714-9

I. Imai: *Applied Hyperfunction Theory.* 1992, 460 pp. ISBN 0-7923-1507-3

Yu.I. Neimark and P.S. Landa: *Stochastic and Chaotic Oscillations.* 1992, 502 pp.
ISBN 0-7923-1530-8

H.M. Srivastava and R.G. Buschman: *Theory and Applications of Convolution Integral Equations.* 1992, 240 pp. ISBN 0-7923-1891-9

Other *Mathematics and Its Applications* titles of interest:

A. van der Burgh and J. Simonis (eds.): *Topics in Engineering Mathematics*. 1992, 266 pp. ISBN 0-7923-2005-3

F. Neuman: *Global Properties of Linear Ordinary Differential Equations*. 1992, 320 pp. ISBN 0-7923-1269-4

A. Dvurecenskij: *Gleason's Theorem and its Applications*. 1992, 334 pp. ISBN 0-7923-1990-7

D.S. Mitrinovic, J.E. Pecaric and A.M. Fink: *Classical and New Inequalities in Analysis*. 1992, 740 pp. ISBN 0-7923-2064-6

H.M. Hapaev: *Averaging in Stability Theory*. 1992, 280 pp. ISBN 0-7923-1581-2

S. Gindinkin and L.R. Volevich: *The Method of Newton's Polyhedron in the Theory of PDE's*. 1992, 276 pp. ISBN 0-7923-2037-9

Yu.A. Mitropolsky, A.M. Samoilenko and D.I. Martinyuk: *Systems of Evolution Equations with Periodic and Quasiperiodic Coefficients*. 1992, 280 pp. ISBN 0-7923-2054-9

I.T. Kiguradze and T.A. Chanturia: *Asymptotic Properties of Solutions of Nonautonomous Ordinary Differential Equations*. 1992, 332 pp. ISBN 0-7923-2059-X

V.L. Kocic and G. Ladas: *Global Behavior of Nonlinear Difference Equations of Higher Order with Applications*. 1993, 228 pp. ISBN 0-7923-2286-X

S. Levendorskii: *Degenerate Elliptic Equations*. 1993, 445 pp. ISBN 0-7923-2305-X

D. Mitrinovic and J.D. Kečkić: *The Cauchy Method of Residues, Volume 2*. Theory and Applications. 1993, 202 pp. ISBN 0-7923-2311-8

R.P. Agarwal and P.J.Y Wong: *Error Inequalities in Polynomial Interpolation and Their Applications*. 1993, 376 pp. ISBN 0-7923-2337-8

A.G. Butkovskiy and L.M. Pustyl'nikov (eds.): *Characteristics of Distributed-Parameter Systems*. 1993, 386 pp. ISBN 0-7923-2499-4

B. Sternin and V. Shatalov: *Differential Equations on Complex Manifolds*. 1994, 504 pp. ISBN 0-7923-2710-1

S.B. Yakubovich and Y.F. Luchko: *The Hypergeometric Approach to Integral Transforms and Convolutions*. 1994, 324 pp. ISBN 0-7923-2856-6

C. Gu, X. Ding and C.-C. Yang: *Partial Differential Equations in China*. 1994, 181 pp. ISBN 0-7923-2857-4

V.G. Kravchenko and G.S. Litvinchuk: *Introduction to the Theory of Singular Integral Operators with Shift*. 1994, 288 pp. ISBN 0-7923-2864-7

A. Cuyt (ed.): *Nonlinear Numerical Methods and Rational Approximation II*. 1994, 446 pp. ISBN 0-7923-2967-8

Other *Mathematics and Its Applications* titles of interest:

G. Gaeta: *Nonlinear Symmetries and Nonlinear Equations*. 1994, 258 pp. ISBN 0-7923-3048-X

V.A. Vassiliev: *Ramified Integrals, Singularities and Lacunas*. 1995, 289 pp. ISBN 0-7923-3193-1

N.Ja. Vilenkin and A.U. Klimyk: *Representation of Lie Groups and Special Functions*. Recent Advances. 1995, 497 pp. ISBN 0-7923-3210-5

Yu. A. Mitropolsky and A.K. Lopatin: *Nonlinear Mechanics, Groups and Symmetry*. 1995, 388 pp. ISBN 0-7923-3339-X

R.P. Agarwal and P.Y.H. Pang: *Opial Inequalities with Applications in Differential and Difference Equations*. 1995, 393 pp. ISBN 0-7923-3365-9

A.G. Kusraev and S.S. Kutateladze: *Subdifferentials: Theory and Applications*. 1995, 408 pp. ISBN 0-7923-3389-6

M. Cheng, D.-G. Deng, S. Gong and C.-C. Yang (eds.): *Harmonic Analysis in China*. 1995, 318 pp. ISBN 0-7923-3566-X

M.S. Livšic, N. Kravitsky, A.S. Markus and V. Vinnikov: *Theory of Commuting Nonselfadjoint Operators*. 1995, 314 pp. ISBN 0-7923-3588-0

A.I. Stepanets: *Classification and Approximation of Periodic Functions*. 1995, 360 pp. ISBN 0-7923-3603-8

C.-G. Ambrozie and F.-H. Vasilescu: *Banach Space Complexes*. 1995, 205 pp. ISBN 0-7923-3630-5

E. Pap: *Null-Additive Set Functions*. 1995, 312 pp. ISBN 0-7923-3658-5

C.J. Colbourn and E.S. Mahmoodian (eds.): *Combinatorics Advances*. 1995, 338 pp. ISBN 0-7923-3574-0

V.G. Danilov, V.P. Maslov and K.A. Volosov: *Mathematical Modelling of Heat and Mass Transfer Processes*. 1995, 330 pp. ISBN 0-7923-3789-1

A. Laurinčikas: *Limit Theorems for the Riemann Zeta-Function*. 1996, 312 pp. ISBN 0-7923-3824-3

A. Kuzhel: *Characteristic Functions and Models of Nonself-Adjoint Operators*. 1996, 283 pp. ISBN 0-7923-3879-0

G.A. Leonov, I.M. Burkin and A.I. Shepeljavyi: *Frequency Methods in Oscillation Theory*. 1996, 415 pp. ISBN 0-7923-3896-0

B. Li, S. Wang, S. Yan and C.-C. Yang (eds.): *Functional Analysis in China*. 1996, 390 pp. ISBN 0-7923-3880-4

P.S. Landa: *Nonlinear Oscillations and Waves in Dynamical Systems*. 1996, 554 pp. ISBN 0-7923-3931-2

Other *Mathematics and Its Applications* titles of interest:

A.J. Jerri: *Linear Difference Equations with Discrete Transform Methods*. 1996, 462 pp. ISBN 0-7923-3940-1

I. Novikov and E. Semenov: *Haar Series and Linear Operators*. 1997, 234 pp. ISBN 0-7923-4006-X

L. Zhizhiashvili: *Trigonometric Fourier Series and Their Conjugates*. 1996, 312 pp. ISBN 0-7923-4088-4

R.G. Buschman: *Integral Transformation, Operational Calculus, and Generalized Functions*. 1996, 246 pp. ISBN 0-7923-4183-X

V. Lakshmikantham, S. Sivasundaram and B. Kaymakcalan: *Dynamic Systems on Measure Chains*. 1996, 296 pp. ISBN 0-7923-4116-3

D. Guo, V. Lakshmikantham and X. Liu: *Nonlinear Integral Equations in Abstract Spaces*. 1996, 350 pp. ISBN 0-7923-4144-9

Y. Roitberg: *Elliptic Boundary Value Problems in the Spaces of Distributions*. 1996, 427 pp. ISBN 0-7923-4303-4

Y. Komatu: *Distortion Theorems in Relation to Linear Integral Operators*. 1996, 313 pp. ISBN 0-7923-4304-2

A.G. Chentsov: *Asymptotic Attainability*. 1997, 336 pp. ISBN 0-7923-4302-6